Dislocation Mechanics of Metal Plasticity and Fracturing

Dislocation Mechanics of Metal Plasticity and Fracturing

Editor
Ronald W. Armstrong

MDPI • Basel • Beijing • Wuhan • Barcelona • Belgrade • Manchester • Tokyo • Cluj • Tianjin

Editor
Ronald W. Armstrong
University of Maryland
USA

Editorial Office
MDPI
St. Alban-Anlage 66
4052 Basel, Switzerland

This is a reprint of articles from the Special Issue published online in the open access journal *Metals* (ISSN 2075-4701) (available at: https://www.mdpi.com/journal/metals/special_issues/metal_dislocation).

For citation purposes, cite each article independently as indicated on the article page online and as indicated below:

LastName, A.A.; LastName, B.B.; LastName, C.C. Article Title. *Journal Name* **Year**, *Article Number*, Page Range.

ISBN 978-3-03943-264-6 (Hbk)
ISBN 978-3-03943-265-3 (PDF)

Cover image courtesy of Ronald W. Armstrong.

© 2020 by the authors. Articles in this book are Open Access and distributed under the Creative Commons Attribution (CC BY) license, which allows users to download, copy and build upon published articles, as long as the author and publisher are properly credited, which ensures maximum dissemination and a wider impact of our publications.

The book as a whole is distributed by MDPI under the terms and conditions of the Creative Commons license CC BY-NC-ND.

Contents

About the Editor vii

Preface to "Dislocation Mechanics of Metal Plasticity and Fracturing" ix

Ronald W. Armstrong
Dislocation Mechanics Pile-Up and Thermal Activation Roles in Metal Plasticity and Fracturing
Reprinted from: *Metals* **2019**, *9*, 154, doi:10.3390/met9020154 1

Hiroyuki Yamada, Tsuyoshi Kami and Nagahisa Ogasawara
Effects of Testing Temperature on the Serration Behavior of an Al–Zn–Mg–Cu Alloy with Natural and Artificial Aging in Sharp Indentation
Reprinted from: *Metals* **2020**, *10*, 597, doi:10.3390/met10050597 9

Martin Diehl, Jörn Niehuesbernd and Enrico Bruder
Quantifying the Contribution of Crystallographic Texture and Grain Morphology on the Elastic and Plastic Anisotropy of bcc Steel
Reprinted from: *Metals* **2019**, *9*, 1252, doi:10.3390/met9121252 21

Ali Waqas, Xiansheng Qin, Jiangtao Xiong, Chen Zheng and Hongbo Wang
Analysis of Ductile Fracture Obtained by Charpy Impact Test of a Steel Structure Created by Robot-Assisted GMAW-Based Additive Manufacturing
Reprinted from: *Metals* **2019**, *9*, 1208, doi:10.3390/met9111208 43

Wolfgang Blum, Jiři Dvořák, Petr Král, Philip Eisenlohr and Vaclav Sklenička
Strain Rate Contribution due to Dynamic Recovery of Ultrafine-Grained Cu–Zr as Evidenced by Load Reductions during Quasi-Stationary Deformation at $0.5\,T_m$
Reprinted from: *Metals* **2019**, *9*, 1150, doi:10.3390/met9111150 55

Wolfgang Blum, Jiři Dvořák, Petr Král, Philip Eisenlohr and Vaclav Sklenička
Quasi-Stationary Strength of ECAP-Processed Cu-Zr at $0.5\,T_m$
Reprinted from: *Metals* **2019**, *9*, 1149, doi:10.3390/met9111149 73

Haochun Tang, Tso-Fu Mark Chang, Yaw-Wang Chai, Chun-Yi Chen, Takashi Nagoshi, Daisuke Yamane, Hiroyuki Ito, Katsuyuki Machida, Kazuya Masu and Masato Sone
Nanoscale Hierarchical Structure of Twins in Nanograins Embedded with Twins and the Strengthening Effect
Reprinted from: *Metals* **2019**, *9*, 987, doi:10.3390/met9090987 87

Hengxu Song and Stefanos Papanikolaou
From Statistical Correlations to Stochasticity and Size Effects in Sub-Micron Crystal Plasticity
Reprinted from: *Metals* **2019**, *9*, 835, doi:10.3390/met9080835 97

Yinan Cui, and Nasr Ghoniem
Influence of Size on the Fractal Dimension of Dislocation Microstructure
Reprinted from: *Metals* **2019**, *9*, 478, doi:10.3390/met9040478 107

Kanji Ono
Size Effects of High Strength Steel Wires
Reprinted from: *Metals* **2019**, *9*, 240, doi:10.3390/met9020240 117

Chandra S. Pande and Ramasis Goswami
Dislocation Emission and Crack Dislocation Interactions
Reprinted from: *Metals* **2020**, *10*, 473, doi:10.3390/met10040473 137

A. Toshimitsu Yokobori, Jr.
Holistic Approach on the Research of Yielding, Creep and Fatigue Crack Growth Rate of Metals Based on Simplified Model of Dislocation Group Dynamics
Reprinted from: *Metals* **2020**, *10*, 1048, doi:10.3390/met10081048 151

About the Editor

Ronald W. Armstrong (Professor Emeritus) obtained a Bachelor of Engineering Science (BES) degree from the Johns Hopkins University in 1955 and a Doctor of Philosophy (PhD) degree in metallurgical engineering from Carnegie Institute of Technology, now within Carnegie-Mellon University, in 1958. A post-doctoral year was spent at the Houldsworth School of Applied Science, Leeds University, UK, in 1958-9, followed by a Westinghouse Research Laboratory appointment in 1959-64, and then at the Commonwealth Scientific and Industrial Research Organization (CSIRO), Division of Tribophysics, University of Melbourne, Australia, in 1964. His academic positions were at Brown University, 1965-68, and the University of Maryland, College Park, 1968-1999. From 2000-2003, he was a senior scientist at the Munitions Directorate, Eglin Air Force Base, FL. Temporary positions have been with the US Office of Naval Research, London, UK, in 1982-84 and 1991, as a liaison scientist, and at other US government and overseas university and government laboratories. His research experience has mainly dealt with the dislocation mechanics of plasticity and fracturing in polycrystalline materials.

Preface to "Dislocation Mechanics of Metal Plasticity and Fracturing"

We begin with a historical description beginning at the start of the 20th century, with a new focus on the effect of surface steps, notches, internal holes, and cracks in reducing the strength of engineering structures. In that first decade, Inglis reported pioneering mechanics calculations of the strength reduction produced by sharp notches and cracks; and, in the second decade, Griffith produced an inverse square root of crack size prediction for the fracture stress of a pre-cracked material. In the third decade, new model analyses were reported of smaller, atomic-scale "dislocation" defects determining the (permanent) plastic deformation behaviors of crystalline materials. That such dislocation defects in localized slip band "pile-ups" behave similarly to Griffith cracks on a continuum mechanics level was described both theoretically and experimentally at the beginning of the fifth decade.

In a complementary manner, the use of optical reflection microscopy to study the crystal microstructures of sectioned metal surfaces was developed by Sorby just before the beginning of the 20th century, and the follow-on discovery of crystal X-ray diffraction in the first decade of the new century led, by the middle of the 20th century, to the use of transmission electron microscopy for observing dislocations in deformed metal foils and, subsequently, to the multiple electron microscope methods that are employed today in modern research investigations probing beneath the surfaces of all types of crystals, almost all of which are full of dislocations. Such dislocations played a counterpart-biological "nematode" role in underlying the new 20th century subject of "Materials Science and Engineering".

Building onto such historical descriptions, the aim of the present *Metals* Special Issue is to provide a valuable sampling of updated research reports focusing on the strength and/or fracturing properties of a variety of modern engineering metals and their alloys. In the introductory article is given a description, based on dislocation mechanics, of the influences of polycrystalline grain size on the hardness, yield stress, and fracture stress of metals and alloys, and which influences are related to an analogously associated crack size dependence. The subsequent all-important research articles begin with a report on the serrated plastic stress–strain behavior exhibited in an aluminum–zinc–magnesium–copper alloy and analyzed in terms of mobile dislocation and atomic solute interactions. Then comes a report on crystallographic grain textures associated with elastic anisotropy measurements in steel materials, followed by an article on the evaluation of Charpy impact test measurements employed to evaluate steel loading rate and notch sensitivity dependencies. Tandem reports are given on dislocation-based assessments of severely deformed copper–zirconium alloy material strength dependencies on applied loading rate. Next, a computational model simulation is described at microscale dimensions of deformation twinning and detwinning in nanograin-sized gold–copper alloy crystals. Two reports follow: first, on the statistical aspects of dislocations tracked in small (fcc) crystal micropillars and, then, on fractal characterizations of dislocations relating to (bcc) iron micropillar test specimens. At the opposite dimensional scale, Weibull characterization of the strength of steel wires as employed in transportation-based bridge cables is reported. This subject connects with the next report on a fracture mechanics description of crack tip plasticity. Lastly, a holistic description is given of the dynamics of dislocation pile-ups in iron and steel materials as related to plastic yielding behavior, creep, and fatigue crack growth rate results. This Special Issue project has been an informative and appreciative effort for me as Guest

Editor. Sincere thanks are expressed to the authors, reviewers, and especially to Ms. Maggie Guo for their super efforts in producing the present Special Issue.

Ronald W. Armstrong
Editor

Editorial

Dislocation Mechanics Pile-Up and Thermal Activation Roles in Metal Plasticity and Fracturing

Ronald W. Armstrong

Center for Engineering Concepts Development, Department of Mechanical Engineering,
University of Maryland, College Park, MD 20742, USA; rona@umd.edu; Tel.: +1-410-723-4616

Received: 14 January 2019; Accepted: 27 January 2019; Published: 31 January 2019

Abstract: Dislocation pile-up and thermal activation influences on the deformation and fracturing behaviors of polycrystalline metals are briefly reviewed, as examples of dislocation mechanics applications to understanding mechanical properties. To start, a reciprocal square root of grain size dependence was demonstrated for historical hardness measurements reported for cartridge brass, in line with a similar Hall-Petch grain size characterization of stress-strain measurements made on conventional grain size and nano-polycrystalline copper, nickel, and aluminum materials. Additional influences of loading rate (and temperature) were shown to be included in a dislocation model thermal activation basis, for calculated deformation shapes of impacted solid cylinders of copper and Armco iron materials. Connection was established for such grain size, temperature, and strain rate influences on the brittle fracturing transition exhibited by steel and other related metals. Lastly, for AISI 1040 steel material, a fracture mechanics based failure stress dependence on the inverse square root of crack size was shown to approach the yield stress at a very small crack size, also in line with a Hall-Petch dependence of the stress intensity on polycrystal grain size.

Keywords: dislocation mechanics; yield strength; grain size; thermal activation; strain rate; impact tests; brittleness transition; fracturing; crack size; fracture mechanics

---

1. Historical Background Leading to Dislocation Mechanics

A substantial improvement to the strength properties of metals by means of refining their internal crystal or grain size has been known for centuries [1]. An example is shown in Figure 1 of early 20th century measurements on the Brinell Hardness Number (BHN) of alpha brass materials, being shown in later work to follow a reciprocal square root of grain diameter dependence [2].

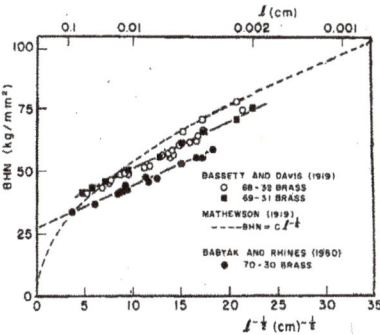

Figure 1. The Brinell Hardness Number (BHN) of alpha brass materials as a function of the reciprocal square root of polycrystalline grain size, $\ell^{-1/2}$; see Reference [2] for references, 1.0 kg/mm^2 = 9.81 MPa.

In Figure 1, the BHN is defined by the load applied to a ball indenter divided by the surface area of the (assumed) spherical cap of the residual permanent indentation. The hardness is more often specified as a Meyer hardness (MH), for which the projected surface area of the residual indent is employed. The MH would show essentially the same grain size dependence as has now been established for many metals and alloys.

The hardness relates to the compressive or tensile flow stress at true strain, ε, through the relationship: MH $\approx$ 3 σ_ε and ε = ~7.5%. Thus, explanation of a grain size dependent hardness follows from association of the hardness with a unidirectional flow stress in which grain size dependence, known as a Hall-Petch relationship, has been explained in terms of dislocation pile-ups in slip bands behaving similarly to shear cracks when blocked at grain boundaries [3]:

$$\sigma_\varepsilon = \sigma_{0\varepsilon} + k_\varepsilon \ell^{-1/2} \tag{1}$$

In Equation (1), $\sigma_{0\varepsilon}$ is the ordinate axis intercept stress taken to apply for plastic flow within the grain volume, k_ε is the microstructural stress intensity required for overcoming the grain boundary resistance, and ℓ is average grain diameter generally measured by a line intercept method. Very interestingly in Figure 1, the prominent metallurgist, Champion Mathewson, proposed that the hardness measurements could be approximated by an $\ell^{-1/4}$ dependence if the hardness was required to be zero for a single crystal. Otherwise, the finite ordinate intercept, $\sigma_{0\varepsilon}$, has been correlated in a number of cases with single crystal plastic flow stress measurements.

2. Nanopolycrystal Hall-Petch Grain Size Strengthening

Many experimental and theoretical investigations have been reported on H-P dependence [4]. Current interest centers on the achievement of an order of magnitude increase in yield strength, that is achieved at nano-scale grain size dimensions. Figure 2 provides a log/log representation of conventional and nano-polycrystalline H-P measurements following Equation (1) for copper, nickel, and aluminum, also at different values of strain [5]. An approximate order of magnitude grain size strengthening effect is observed. The indicated low $k_{0.14}$ value for nickel at large strain and conventional grain sizes, compared to the nano-scale k_ε obtained from hardness measurements, is an anomalous result. At smaller proof strains, nickel k_ε has been shown to be near to that of copper.

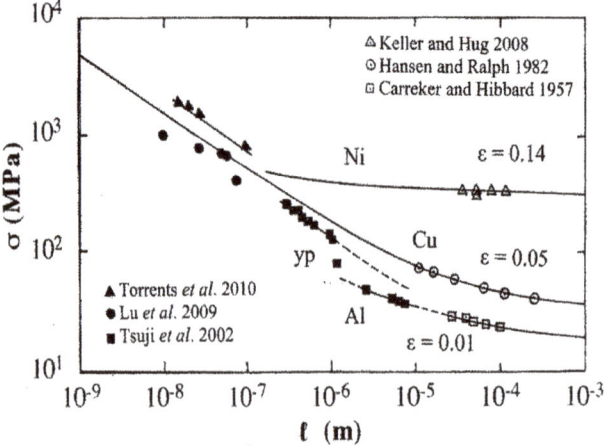

Figure 2. Comparison of Hall-Petch grain size dependent strengthening results at conventional and nano-polycrystalline grain sizes for Al, Cu, and Ni materials; the referenced data are given in Reference [5].

The near-equivalence of copper and nickel k_ε values, along with a significantly lower value for aluminum, has been explained in terms of the calculated shear stress at the pile-up tip, τ_C, being correlated with the need for cross-slip in effecting transmission of plastic flow across grain boundaries [5]; see Equation (2):

$$k_\varepsilon = m_T[\pi m_S G b \tau_C / 2\alpha']^{1/2} \quad (2)$$

In Equation (2), m_T and m_S are Taylor and Sachs orientation factors, G is shear modulus, b is Burgers vector, and $\alpha' \approx 0.8$ is for an average dislocation character. A nearly equivalent numerical value of $Gb\tau_C$ is obtained for copper and nickel, thus explaining their nearly same k_ε values and, also, is consistent for aluminum with observation of a much lower k_ε value controlled by easy cross-slip. The indicated increase in k_ε in Figure 2 for aluminum when exhibiting a well-defined yield point, (yp), is correlated also with the well-established measurement of a much larger k_{yp} for yield point behavior, for example, in steel.

3. Thermally-Activated Dislocation Mobility Relations

Jeffries reported in 1919 pioneering measurements on the combination of grain size, temperature, and strain rate dependencies of the mechanical properties of annealed and deformed copper materials [6]. A considerable number of other reports have followed on the topic, particularly involving the deformation of metal single crystals first produced during the same beginning period of the 20th century. Seeger reported in 1958 a summary description of fcc crystal deformation properties in terms of thermally-activated dislocation motion [7]. The report was followed in 1973 by the inclusion of an H-P dependence for polycrystals [8]. The single crystal/polycrystal topic was reviewed in 2008, with emphasis given to constitutive relations developed for deformation dynamics calculations under condition of high rate loading [9].

3.1. Thermally-Activated FCC Strain Hardening

The thermal dependence is in the strain hardening, $d\sigma_\varepsilon/d\varepsilon$, for fcc metals and alloys. One of several dislocation mechanics based constitutive equations proposed for σ_ε is given by [10]:

$$\sigma_\varepsilon = \sigma_{G\varepsilon} + B_0\{\varepsilon_r \cdot [1 - \exp(-\varepsilon/\varepsilon_r)]\}^{1/2} \exp(-\alpha T) + k_\varepsilon \ell^{-1/2} \quad (3)$$

In Equation (3), $\sigma_{G\varepsilon}$ is an athermal stress for elastic interactions within the polycrystal grain volumes, ε_r is a reference strain for dynamic recovery, and $\alpha = \alpha_1 - \alpha_2 \ln(d\varepsilon/dt)$ is a temperature coefficient including strain rate, $(d\varepsilon/dt)$, dependence that is rooted in the thermally-activated dislocation rate description. The first two terms are included within $\sigma_{0\varepsilon}$ in Equation (1). At small ε values, σ_ε follows a parabolic Taylor-type strain dependence.

An example calculation employing Equation (3) to describe the deformation shape of an impacted solid cylinder in comparison with the experimental shape is shown in Figure 3. The computed deformation shape, obtained with use of separately determined material constants from reference stress–strain tests, was achieved with the Elastic Plastic Impact Calculation (EPIC) code [11]. A slight improvement in the calculated deformation profile was obtained over another calculation applied to the same test result employing the eponymous Johnson-Cook numerical equation developed jointly with invention of the EPIC code.

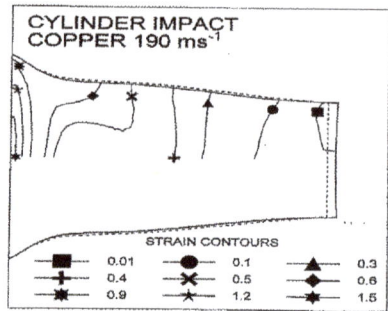

Figure 3. Comparison of calculated (continuous curve) and (dotted) experimental shapes for a longitudinal section of an impacted copper solid cylinder, including internal iso-strain profiles [10].

3.2. Thermal BCC Yield Stress

For bcc metals and alloys, the thermal dependence is in the lower yield point stress, $\sigma_{lyp} = \sigma_\varepsilon$, and the strain hardening is athermal. The counterpart constitutive equation for the behavior is given in Reference [10]:

$$\sigma_\varepsilon = \sigma_{G\varepsilon} + B\exp(-\beta T) + A\varepsilon^n + k_\varepsilon \ell^{-1/2} \qquad (4)$$

In Equation (4), $\beta = \beta_0 - \beta_1 \ln(d\varepsilon/dt)$ and A and n are experimental constants describing a power law dependence for the strain hardening and the other parameters are the same as defined in Equation (3). Figure 4 shows an example deformation shape for an Armco iron solid cylinder impacted in the same manner as was done for copper in Figure 3, which the result has included the additional complication of deformation by twinning in the early stage of impact [11].

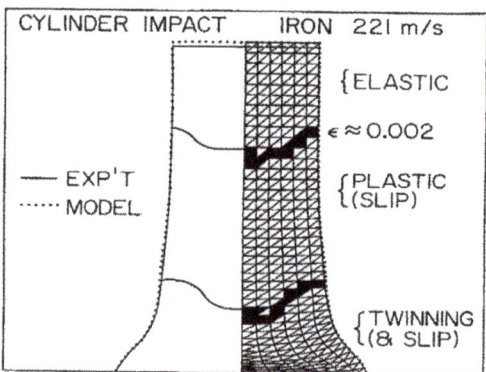

Figure 4. Experimental and Elastic Plastic Impact Calculation (EPIC) modeled solid cylinder impact test result on Armco iron as a result of initial athermal deformation twinning then followed by thermally-activated slip [10].

A previous report on solid cylinder impact tests made on α-iron material had revealed in the region close to the impact surface the occurrence of deformation twins, then called Neumann bands, after their observation in meteorites [12]. Sequential EPIC calculations applied to the result shown in Figure 4 revealed that a limited amount of essentially athermal twinning occurred first on impact and hardened the material, in part, by grain size reduction, and then further deformation followed afterward by thermally-activated slip. Such twinning is known to follow an H-P dependence with constants, $\sigma_{0T} < \sigma_{0\varepsilon}$ and $k_T > k_\varepsilon$, thus indicating a transition at smaller grain size when total deformation by slip is preferred to twinning. The profile of Figure 4 was shown to be essentially identical to the originally reported longitudinal section view containing the Neumann band structure [10,12].

4. Brittle Fracturing and Fracture Mechanics

At lower temperatures or higher plastic strain rates, brittle cleavage fracturing intervenes in tensile tests of steel and related bcc metals and alloys. The tensile cleavage fracture stress also follows an H-P dependence with a higher value of $k_C > k_T$. The characteristic temperature, T_C, for the transition in behavior has been modeled on a dislocation mechanics basis [13]. The topic also relates importantly to the sudden onset of brittle failure that may occur due to the presence of a sharp crack, as included in the subject of fracture mechanics.

4.1. The Ductile-to-Brittle Transition Temperature (DBTT)

The brittleness transition behavior is depicted in Figure 5 for a compilation of measurements made on two steel materials with different grain sizes, and including measurements made of tensile yield stress, brittle fracture stresses in bend tests, and Charpy v-notch impact energy tests [13]. In the figure, the effective yield stress in the Charpy test has been raised by a notch factor, $\alpha = 1.94$, to take account of the influence of hydrostatic component of stress and a small value of β has been employed (appropriate to an effective strain rate of 400 s^{-1}); see Equation (4). The effective H-P k_{lyp} associated with the difference in yield stresses for the two grain sizes is seen to be a much smaller effect than the corresponding larger effect of k_C on the fracture stress, σ_C, so producing a lower value of transition temperature for the smaller grain size material. The predicted transition for the smaller grain size was found to be raised somewhat because of easier cracking associated with the presence of carbide plates at the grain boundaries.

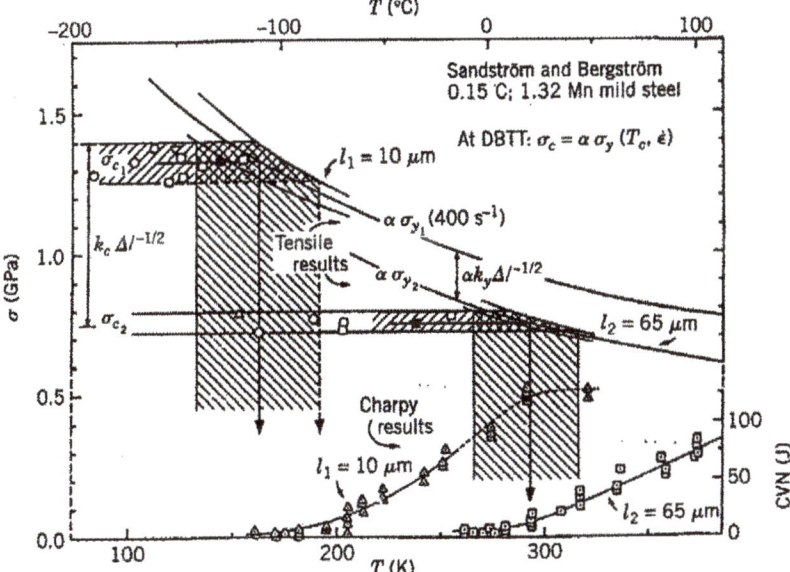

Figure 5. The ductile-to-brittle transition temperatures for two steel materials with different grain sizes of 65 and 10 μm as determined in tensile tests and via Charpy v-notch (CVN) impact tests [13].

4.2. Plastic Zone and Grain Size in Fracture Mechanics

The notch effect in relatively small-scale Charpy impact tests relates to the role of crack size in fracture mechanics tests at micro- to macro-scale dimensions, and to the progression from Griffith's pioneering work on a reciprocal square root of crack size dependence for the fracture stress as extended on a continuum mechanics basis by Irwin [14]. The importance of the plastic zone size in the Charpy test is not obviated by any cracks, no matter how sharp, which are able to be put into a fracture mechanics test specimen [15]. Bilby, Cottrell, and Swinden employed a continuum dislocation pile-up model both for a crack and strip-type plastic zone at the crack tip [16]. The transcendental equation obtained for critical growth of the crack was shown to be closely approximated by the relationship [17]:

$$\sigma_F = A\sigma_y[s/(c+s)]^{1/2} \tag{5}$$

In Equation (5), A is a numerical constant near unity, c is the half-length of an internal crack, and s is the length of the plastic strip. Figure 6 shows application of the relationship with $A = 1.0$ to measurements recently reported for the American Society for Testing and Materials (ASTM) specified fracture mechanics measurements made on AISI 1040 steel material [18].

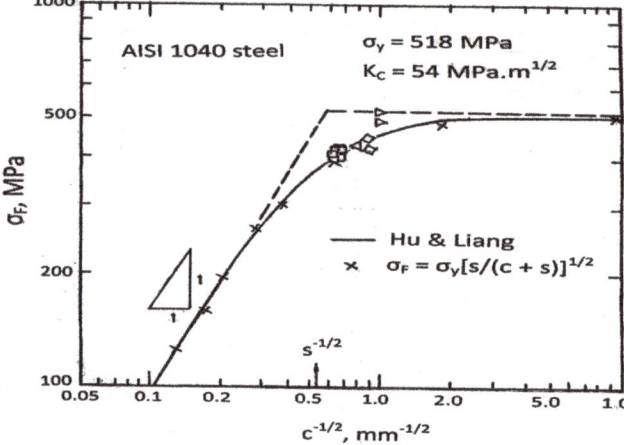

Figure 6. Comparison of fracture mechanics specified stress dependence on crack size for AISI 1040 steel obtained from results reported by Hu and Liang and matched with calculated plastic zone, s.

The extended continuous curve shown in Figure 6 was established by Hu and Liang for an AISI 1040 plate material of length 40 mm including crack size to length ratios between 0.1 and 0.7, and with the separate measurements indicated for the yield stress and plane strain determined stress intensity, K_{IC}, value. At large crack size, a linear dependence of fracture stress, σ_F, on the reciprocal square root of crack size is obtained, as predicted. The indicated fit of x-marked points on the curve were obtained with Equation (5), with $s = 3.5$ mm that corresponds in the Hu and Liang calculation of a critical reference crack length of 4.04 mm. A deviation from the predicted linear fracture mechanics relationship is seen to occur at relatively smaller $(c/s) < \sim 3$ than might have been expected, for example, proceeding onward from $\sigma_F \geq 0.5\,\sigma_y$.

The yield stress dependence in the fracture mechanics description has been extended in the same type analysis to description of the fracture mechanics stress intensity dependence on grain size in the relationship:

$$K_{IC} = (8/3\pi)^{1/2}[\sigma_{0C} + k_C \ell^{-1/2}]s^{1/2} \tag{6}$$

In many cases, s is relatively constant and therefore K_{IC} follows an H-P type dependence [15].

5. Discussion

The examples given of dislocation mechanics based relationships for hardness, fcc and bcc plastic flow stresses, impact, and fracturing properties constitute only a relatively limited number of connections being currently researched for a wide variety of metals and their alloys. An example is provided by hardness measurements being investigated in micro- and nano-indentation tests of nano-dimensional grain size nickel materials [5], and such measurements are being extended to indentation fracture mechanics measurements made on relatively more brittle materials [19].

In addition, the very positive influence of grain size on strengthening metals is being investigated in terms of a variety of material processing methods, in particular, by the method of severe plastic deformation (SPD) [20]. Both an increase in $\sigma_{0\varepsilon}$ and decrease in ℓ contribute to increasing σ_ε. The method has historical connection with wire drawing of patented (eutectoid) steel wire known as piano wire, and which material is more recently being employed in the strengthening of automotive tires at nano-scale iron and iron carbide phase separations [21].

For fcc and hcp metals, there is an important magnification of the plastic strain rate sensitivity measured at nano-scale dimensions in that the pile-up stress, τ_C, in Equation (2) is sufficiently small as to be affected by thermal activation, thus producing a grain size dependence for the activation area, $A^* = v^*/b = (k_B T/b)(\Delta \ln[d\varepsilon/dt]/\Delta \tau_{Th})_T$ in which v^* is the frequently employed activation volume, k_B is the Boltzmann constant, and $\tau_{Th} = \sigma_{Th}/m_T$ is strain dependent. As a consequence, $(1/v^*)$ follows a H-P type dependence:

$$(1/v^*) = (1/v^*_0) + (k_\varepsilon/2m_T\tau_C v^*_C)\ell^{-1/2} \qquad (7)$$

In Equation (7), $(1/v^*_0)$ applies to strain rate sensitivity within the polycrystal grain volumes, and $\tau_C v^*_C \approx \tau_{CTh} v^*_C$ is constant [5]. At very small grain sizes, say <20 nm, there is a reversal in the H-P dependence but the value of v^* is substantially decreased even more to a size of atomic dimensions, coincident with grain size weakening attributed to atomic diffusion mechanisms.

Important strain rate sensitivity is involved also in the ductile–brittle transition behavior described in connection with Figure 5, as is true for the important influence of grain size dependence. However, greater emphasis is given normally to specifying as accurately as possible the fracture mechanics stress intensity parameter, K_{IC}, employed to characterize the propensity of the material for the sudden onset of catastrophic failure. In this case, hardness testing again provides a useful method of characterizing the indentation fracture mechanics properties of relatively more brittle materials [19].

6. Summary

A brief description has been given of hardness, grain size, flow stress, temperature, strain rate, and crack size aspects of dislocation mechanics based descriptions of metal plasticity and fracturing. The purpose of the description has been to provide several examples among the many investigations already reported or being underway, in order to characterize the corresponding mechanical properties of metals and their alloys.

Conflicts of Interest: The author declares no conflict of interest.

References

1. Armstrong, R.W. Plasticity: Grain size effects III. In *Reference Module in Materials Science and Engineering*; Hashmi, S., Ed.; Elsevier: New York City, NY, USA, 2018; pp. 1–23.
2. Jindal, P.C.; Armstrong, R.W. The dependence of the hardness of cartridge brass on grain size. *Trans. TMS-AIME* **1967**, *239*, 1856–1857.
3. Armstrong, R.W. The influence of polycrystal grain size on several mechanical properties. *Metall. Trans.* **1970**, *1*, 1169–1176. [CrossRef]
4. Armstrong, R.W. 60 years of Hall-Petch: Past to present nano-scale connections. *Mater. Trans.* **2014**, *55*, 2–12. [CrossRef]
5. Armstrong, R.W. Hall-Petch description of nanopolycrystalline Cu, Ni and Al strength levels and strain rate sensitivities. *Phil. Mag.* **2016**, *96*, 3097–3108. [CrossRef]
6. Jeffries, Z. Effect of temperature, deformation and grain size on the mechanical properties of metals. *Trans. TMS-AIME* **1919**, *60*, 474–576, with discussion by C.H. Mathewson and others.

7. Seeger, A. Kristallplastizität. In *Handbuch der Physik VII/2, Crystal Physics II*; Flugge, S., Ed.; Springer: Berlin, Germany, 1958; pp. 1–210.
8. Armstrong, R.W. Thermal activation–strain rate analysis (TASRA) for polycrystalline materials. *J. Sci. Ind. Res.* **1973**, *32*, 591–598.
9. Armstrong, R.W.; Walley, S.M. High strain rate properties of metals and alloys. *Intern. Mater. Rev.* **2008**, *53*, 105–128. [CrossRef]
10. Zerilli, F.J.; Armstrong, R.W. Dislocation mechanics based analysis of material dynamics behavior: Enhanced ductility, deformation twinning, shock deformation, shear instability, dynamic recovery. *J. Phys. IV France Colloq.* **1997**, *7*, 637–642. [CrossRef]
11. Johnson, G.R.; Cook, W.H. A constitutive model and data for metals subjected to large strains, high strain rates, and high temperatures. In Proceedings of the 7th International Symposium on Ballistics, The Hague, The Netherlands, 19–21 April 1983; pp. 541–547.
12. Carrington, W.E.; Gaylor, M.L.V. The use of flat-ended projectiles for determining dynamic yield stress III. Changes in microstructure caused by deformation under impact at high striking velocities. *Proc. R. Soc. Lond. A* **1948**, *194*, 323–331.
13. Armstrong, R.W. Material grain size and crack size influences on cleavage fracturing. *Phil. Trans. R. Soc. A* **2015**, *373*, 20140124. [CrossRef] [PubMed]
14. Irwin, G.R. Fracture. In *Handbuch der Physik VI*; Flugge, S., Ed.; Springer: Berlin, Germany, 1958; pp. 551–590.
15. Armstrong, R.W. Crack Size and Grain Size Dependence of the Brittle Fracture Stress. In *Dritte Intern. Tagung uber den Bruck, ICF3*; Kochendorfer, A., Ed.; Verein Deutscher Eisenhuttenleute: Dusseldorf, Germany, 1973; p. III-421.
16. Bilby, B.A.; Cottrell, A.H.; Swinden, K.H. The Spread of Plastic Yield from a Notch. *Proc. R. Soc. Lond. A* **1963**, *272*, 304–314.
17. Armstrong, R.W. Dislocation viscoplasticity aspects of material fracturing. *Eng. Fract. Mech.* **2010**, *77*, 1348–1359. [CrossRef]
18. Hu, X.-Z.; Liang, L. Elastic-Plastic and Quasi-Brittle Fracture. In *Handbook of Mechanics of Materials*; Hsueh, C.H., Schmauder, S., Chen, C.-S., Chawla, K.K., Chawla, N., Chen, W., Kagawa, Y., Eds.; Springer: Singapore, 2019; pp. 1–32, see Figure 11.
19. Armstrong, R.W.; Walley, S.M.; Elban, W.L. Elastic, plastic and cracking aspects of the hardness of materials. *Int. J. Mod. Phys. B* **2013**, *28*, 1330004. [CrossRef]
20. Vinogradov, A.; Estrin, Y. Analytical and numerical approaches to modelling severe plastic deformation. *Prog. Mater. Sci.* **2018**, *95*, 172–242.
21. Armstrong, R.W. Size Effects on Material Yield Strength/Deformation/Fracturing Properties. *J. Mater. Res.* **2019**, in press. [CrossRef]

© 2019 by the author. Licensee MDPI, Basel, Switzerland. This article is an open access article distributed under the terms and conditions of the Creative Commons Attribution (CC BY) license (http://creativecommons.org/licenses/by/4.0/).

Article

Effects of Testing Temperature on the Serration Behavior of an Al–Zn–Mg–Cu Alloy with Natural and Artificial Aging in Sharp Indentation

Hiroyuki Yamada [1,*], Tsuyoshi Kami [2] and Nagahisa Ogasawara [1]

1. Department of Mechanical Engineering, National Defense Academy, 1-10-20 Hashirimizu, Yokosuka-shi 239-8686, Kanagawa, Japan; oga@nda.ac.jp
2. Graduate School of Science and Engineering, National Defense Academy, 1-10-20 Hashirimizu, Yokosuka-shi 239-8686, Kanagawa, Japan; kamimechnda@gmail.com
* Correspondence: ymda@nda.ac.jp; Tel.: +81-46-841-3810

Received: 27 February 2020; Accepted: 30 April 2020; Published: 3 May 2020

Abstract: Serration phenomena, in which stress fluctuates in a saw-tooth shape, occur when a uniaxial test is performed on an aluminum alloy containing a solid solution of solute atoms. The appearance of the serrations is affected by the strain rate and temperature. Indentation tests enable the evaluation of a wide range of strain rates in a single test and are a convenient test method for evaluating serration phenomena. Previously, the serrations caused by indentation at room temperature were clarified using strain rate as an index. In this study, we considered ambient temperature as another possible influential factor. We clarify, through experimentation, the effect of temperature on the serration phenomenon caused by indentation. An Al–Zn–Mg–Cu alloy (7075 aluminum alloy) was used as the specimen. The aging phenomenon was controlled by varying the testing temperature of the solution-treated specimen. Furthermore, the material properties obtained by indentation were evaluated. By varying the testing temperature, the presence and amount of precipitation were controlled and the number of solute atoms was varied. Additionally, the diffusion of solute atoms was controlled by maintaining the displacement during indentations, and a favorable environment for the occurrence of serrations was induced. The obtained results reveal that the variations in the serrations formed in the loading curvature obtained via indentation are attributed to the extent of interaction between the solute atoms and the dislocations.

Keywords: indentation; serration; temperature; strain rate; dislocation; artificial aging; solid solution; loading curvature; aluminum alloy

---

1. Introduction

A phenomenon called serration—stress fluctuations in a saw-tooth shape—occurs when a uniaxial test (e.g., tensile test) is performed on aluminum alloys containing solute atoms in the solid solution [1–4]. Dynamic strain aging gives rise to the Portevin–Le Chatelier effect [3,5,6]. One of the manifestations of this effect is the serration that occurs when dislocations are pinned or released from the atmosphere of a solute atom. For example, in Al–Mg alloys (5000 series aluminum alloys), significant serrations are generated, as well as a strain pattern similar to a Luder's band on the surface of the alloy, thereby impairing its appearance. Many studies on the serration phenomena have been conducted, primarily involving the 5000 series aluminum alloys. Existing studies identified serration behavior as an interaction between dislocations and the solute Mg [3,5,6]. This phenomenon was reported to be affected by strain rate and temperature because the velocity of dislocation motion and the diffusion rate of Mg are affected by the strain rate and the temperature, respectively [3,5].

Previous studies have demonstrated that the serration behavior varies with change in the strain rate. The classifications of serrations are briefly summarized below [7]. Figure 1 shows an example of variations in the serration behavior with a change in strain rate [8]. It shows an A-type stress fluctuation that repeatedly rises and falls (the wavy stress fluctuation at a relatively high strain rate, $\dot{\varepsilon}_A$) and B-type saw-tooth-like stress fluctuations generated at intermediate strain rate ($\dot{\varepsilon}_B$, $\dot{\varepsilon}_{A+B}$). Serrations sometimes occur as a result of a combination of behaviors. For example, an A + B-type—a combination of A and B-type stress—has been reported. The serrations observed at low strain rates ($\dot{\varepsilon}_C$) are classified as C-type stress (the C_A and C_B types depending on the frequency of fluctuation), in which the fluctuations are irregular. At a strain rate above $\dot{\varepsilon}_A$ or below $\dot{\varepsilon}_C$, serrations do not occur, rather, a smooth stress–strain relationship is obtained.

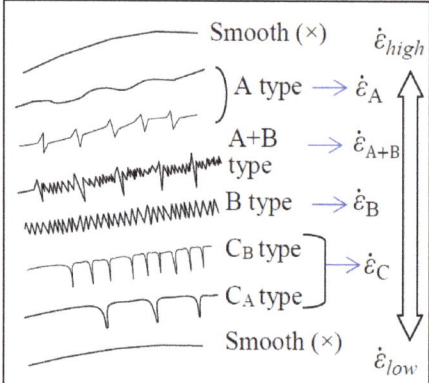

Figure 1. Examples of different serration behaviors for different strain rate regimes [8].

To date, serrations have been evaluated primarily by uniaxial assessments such as compression and tensile tests. The strain rate in the uniaxial test is defined by the following equation:

$$\dot{\varepsilon}_u = \frac{d\varepsilon_u}{dt}, \tag{1}$$

where $\dot{\varepsilon}_u$ is the strain rate and ε_u is the strain during the uniaxial test. However, present authors [8] have, through experiment, proved that serration phenomena can be evaluated through indentation tests. Through continuous measurement of the load and the corresponding displacement, while loading and unloading in an indentation test, some mechanical properties that cannot be obtained through hardness tests could possibly be evaluated [9]. Therefore, the indentation test is also a vital non-destructive test for metals. The mechanical properties obtained through indentation test are evaluated using the loading curvature C, as shown below:

$$P = Ch^2, \tag{2}$$

where P and h are the load and displacement during indentation, respectively. The loading curvature is given by the following relational expression:

$$C = f(E, Y, n, \alpha, T, \dot{\varepsilon}_i), \tag{3}$$

where E is Young's modulus, Y the yield stress, n the work hardening index, α the indenter angle, T the temperature, and $\dot{\varepsilon}_i$ the strain rate of the indentation. Doerner and Nix [10] proposed the following empirical equation for determining the value of $\dot{\varepsilon}_i$ using a triangular pyramid indenter with the same area to depth ratio as the Vickers pyramid:

$$\dot{\varepsilon}_i = k\left(\frac{\dot{h}}{h}\right), \qquad (4)$$

where k is the material constant and $\dot{h}$ is the displacement velocity. This equation does not include the effect of indenter angle. It should be noted here that the concepts of the representative stress (σ_r) and strain (ε_R) for indentation analysis in order to normalize the load–displacement curves [11–13]. In the indentation test, σ_r is the flow stress of the uniaxial test at a particular strain ε_r. The representative strain is given the formula [13]:

$$\varepsilon_R = 0.0638 \cot \alpha, \qquad (5)$$

where α is a half-apex angle. For example, ε_R is approximated as 0.023 when using a conical indenter with a half-apex angle of 70.3° (described later). Therefore, Equation (4) expresses the strain rate at this representative strain value.

The $\dot{\varepsilon}_i$ has the same dimension as $\dot{\varepsilon}_u$, however, the definition of strain rate for indentation tests differs from that of uniaxial tests. The strain rate of indentation is distributed inside the test materials in a complex manner [14]. Thus, existing studies have proposed the concept of effective strain rate, $\dot{\varepsilon}_e$, to consider the effect of the distribution of the strain rate on the indentation [15–17]. The effective strain rate is given by the formula:

$$\dot{\varepsilon}_e = \beta\left(\frac{\dot{h}}{h}\right), \qquad (6)$$

where β is a material constant. Equations (4) and (6) have the same form. However, it has been shown that the value of β correlates the strain rate in indentation with that in uniaxial tests [15–17].

Previous studies [8,18] indicated that the serration phenomenon in indentation could possibly be evaluated using the concept of effective strain rate. Indentation is performed using a sharp indenter, hence, a complicated deformation field is generated in the test material, whose deformation mechanism has only recently been clarified [19]. Through the indentation tests, it was also discovered that there is a test evaluation limit called critical strain [19] and that serrations could be used as an index to evaluate this effect [18]. However, these previous studies were conducted only in an ambient temperature environment, hence, the effect of temperature on the serration behavior during indentation was not investigated. There are many unknown factors that could affect the behavior of the serrations obtained from indentation.

Until now, the strengths of most metals are evaluated through uniaxial tensile tests. However, next-generation metals are expected to have micro- and nano-scale properties. Therefore, there is a need to adopt such tests as an indentation test that can non-destructively evaluate the strength of a small area with accuracy comparable to that of uniaxial tests. In this study, we extend our previous work [8,18] to clarify the effects of testing temperature on the serration behavior during indentation tests. The microstructural changes in the Al–Zn–Mg–Cu alloy (7075 aluminum alloy) due to natural and artificial aging were employed [20]. In addition, indentation was established as a new method of evaluating material properties through the evaluation of the serration behavior related to the microstructure.

2. Materials and Methods

2.1. Specimen

A 7075 aluminum alloy (hereafter referred to as the 7075 alloy) specimen was used in this study. Table 1 lists the chemical composition of the 7075 alloy. The dimensions of the cylindrical specimen were 40 mm (diameter) and 40 mm (height), and the end face was finished by lathing. The solution treatment was conducted at 753 K for 3600 s, followed by water-cooling. Indentation tests were performed on this solution-treated specimen.

Table 1. Chemical composition of the investigated 7075 alloy (wt%).

Alloy	Si	Fe	Cu	Mn	Mg	Cr	Zn	Ti	Al
7075	0.09	0.19	1.6	0.04	2.6	0.20	5.6	0.02	Bal.

2.2. Indentation

2.2.1. Testing Conditions

A universal testing machine (Instron, series 5982, Norwood, MA, USA) attached with a jig was used for the indentation. Approximately 1 mm was loaded into the lathe-machined surface of the specimen as milli-indentation. A conical indenter (Figure 2a) made of a WC–Co superalloy was used. The radius of the tip (referred to as the roundness) of the indenter was 8.63 μm. The indenter angle was measured to be 141.02° using a laser microscope. This angle is almost equal to that of a conical indenter (140.6°) that has the same indentation projection area at the same indentation depth as those of the Berkovich indenter (a triangular pyramid with a ridge angle of 115°; see Figure 2b). In the existing studies, the effects of the roundness of indenters up to the initial stage of indentation (approximately 1 μm) were reported, where the roundness of the indenter was approximately 10 μm. However, the effect was obtained to be negligible when the indentation was higher than 1 μm [21]. In this study, we assumed an indentation of 1 mm. Therefore, the effect of the roundness of the indenter was minimal.

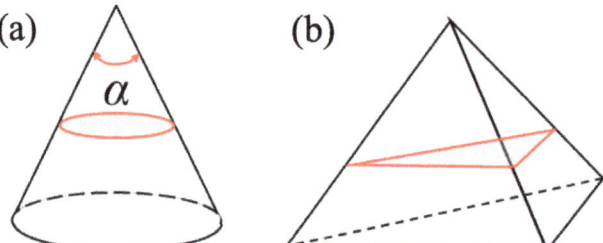

Figure 2. Schematic diagram of a cone-type indenter: (**a**) conical type and (**b**) Berkovich (triangular pyramid) type. Here α is the indenter angle and the shaded section is the projection area.

In this study, the temperature at which the Guinier–Preston (GP) zones and the η' phase precipitate in 7075 alloys was adopted as the testing temperature (described later). First, a temperature of 343 K at which GP zones have been confirmed to precipitate (or nucleate) [20,22] was chosen. The η' phase has often precipitated under the condition of aging at 393 K for 24 h (T6 temper). In this study, however, a temperature of 443 K at which the η' phase has been reported to precipitate within a short time (600 to 3600 s) [20] was adopted. Figure 3 shows a schematic diagram of the indentation test at high temperatures. A home-built electric furnace was used. A thermocouple was attached to the specimen surface and was controlled to a predetermined temperature using a temperature controller (CHINO, SY2111, Tokyo, Japan). The arrival times at 343 and 443 K were approximately 1800 and 3600 s, respectively. The specimen was held for 900 s at each of the temperatures, after which indentation was performed. Furthermore, the specimen was held at 77 K in a liquid nitrogen where the low-temperature indentation was performed. At this temperature, aging takes place at a very low rate. Unlike the high-temperature test, the low-temperature indentation test was performed in a container using waterproof paper. The specimen was also indented at room temperature (293 K), thus, the test was performed in four different temperature environments. For each of the temperature conditions, the test preparation was initiated within 300 s of the solution treatment. Therefore, the effect of natural aging was small, except at room temperature.

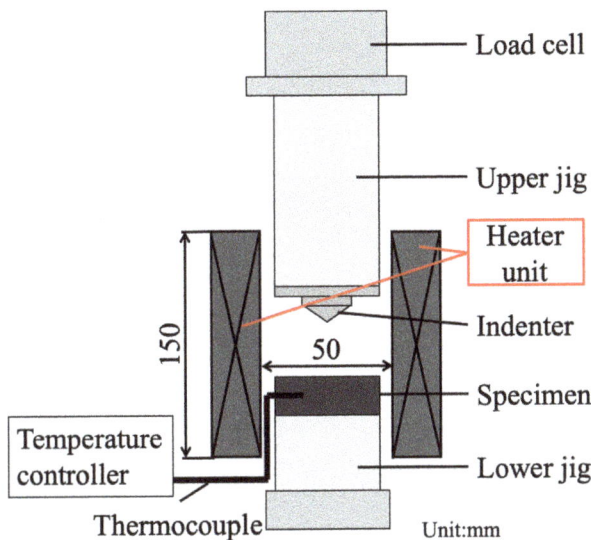

Figure 3. A schematic diagram of milli-indentation at high temperatures (343 and 443 K).

2.2.2. Indenter Control

The control methods for indenters can be classified into two: loading rate and displacement rate controls. In this study, a commercial universal testing machine was used, hence, displacement rate control was employed. During indentation, the strain rate was varied by varying the displacement rate (see Figure 4). Previous studies have shown that serrations are being affected by the diffusion of Mg in the solid solution [5]. The time required for Mg atoms in the solid solution to sufficiently pin stagnant dislocations is given by the following equation [23,24]:

$$t_a = \left(\frac{C_1}{3C_0}\right)^{\frac{3}{2}} \frac{kTb^2}{3DU_m}, \tag{7}$$

where C_1 is the concentration of solid solution atoms required for serrations to occur, C_0 the concentration of solid solution atoms in the material, k the Boltzmann's constant, T temperature, D the diffusion rate, and U_m the binding energy between the solid solution atoms and dislocations. When clusters and GP zones are formed by aging, the amount of Mg in the solid solution decreases, and accordingly, C_0 decreases. D also increases as temperature increases [25,26]. It is difficult to measure the diffusion rate of solute atoms at low temperatures. Therefore, we predicted using the following equation:

$$D = D_0 \exp\left(-\frac{Q}{RT}\right), \tag{8}$$

where D_0 is the frequency factor, Q is the activation energy and R is the gas constant. Table 2 shows the prediction of the diffusion rate of solute atoms in aluminum at the testing temperature [27–29]. Hence, to promote the diffusion of solute atoms, a constant displacement is maintained. As shown in Figure 4, the displacement was held constant at two different values, for 20 s each, during the indentation test.

Table 2. Prediction of the diffusion rate of solute atoms in aluminum at the testing temperature using Equation (8).

Solute Atom	D_0 (m²/s)	Q (kcal/mol)	Testing Temperature (K)	D (m²/s)	Reference
Zn	1.77×10^{-5}	28.0	77	7.74×10^{-85}	[27]
			293	2.47×10^{-26}	
			343	2.71×10^{-23}	
			443	2.85×10^{-19}	
Mg	6.23×10^{-6}	27.5	77	7.11×10^{-84}	[28]
			293	2.05×10^{-26}	
			343	1.98×10^{-23}	
			443	1.77×10^{-19}	
Cu	1.5×10^{-5}	30.2	77	3.81×10^{-91}	[29]
			293	4.81×10^{-28}	
			343	9.14×10^{-25}	
			443	1.99×10^{-20}	

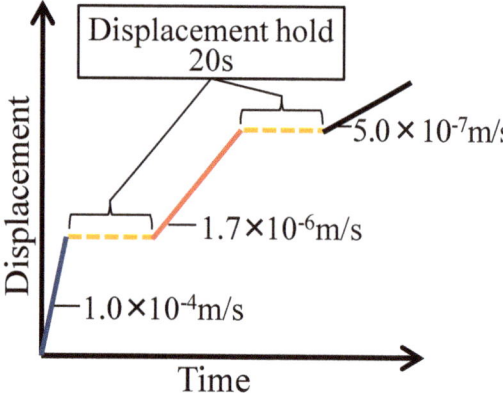

Figure 4. A schematic diagram of the time history of displacement.

3. Results

Figure 5 shows the load–displacement relationships for all the tests at different temperatures. For the tests at temperatures above room temperature, the load increased with an increase in temperature regardless of the change in the displacement rate. This increase in the load is attributed to the precipitates formed by artificial aging. At 77 K, a decrease in the displacement rate lowered the increase rate of the load as the displacement increased as compared to the other temperatures.

The effective strain rates for the indentations were calculated using Equation (6). Herein, $\beta = 0.1$ was used, based on previous studies [8,18]; the effective strain rate–displacement relationship is shown in Figure 6. The effective strain rate under displacement rate control, given by Equation (6), decreased as the displacement increased. The indentations were performed at three different rates by varying the indenter speed (see Figure 4). A wide range of effective strain rates (from 10^{-4} to 10^0 s^{-1}) was obtained during the indentations. There was no significant difference in the effective strain rate, even when the testing temperature changed. Aluminum alloys are known to have high strength–strain rate sensitivity at lower temperatures [30,31]. This implies that at extremely low temperatures, the strength of aluminum alloys decreases as the strain rate decreases. Therefore, during the cryogenic indentation, the decrease in the increase rate of the load was as a result of the decrease in the effective strain rate with increasing displacement.

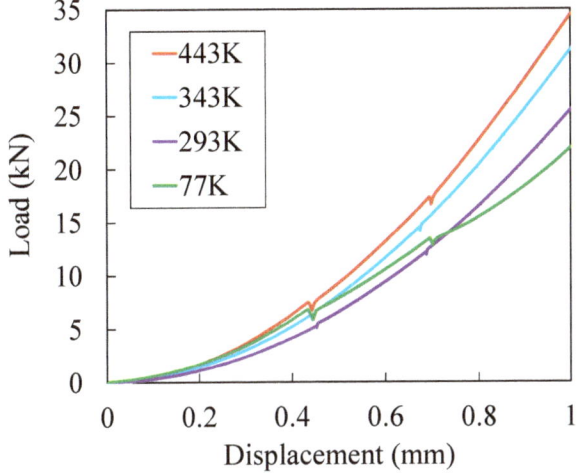

Figure 5. Load–displacement curves at 77, 293, 343, and 443 K. The change in displacement rate during indentation is shown in Figure 4.

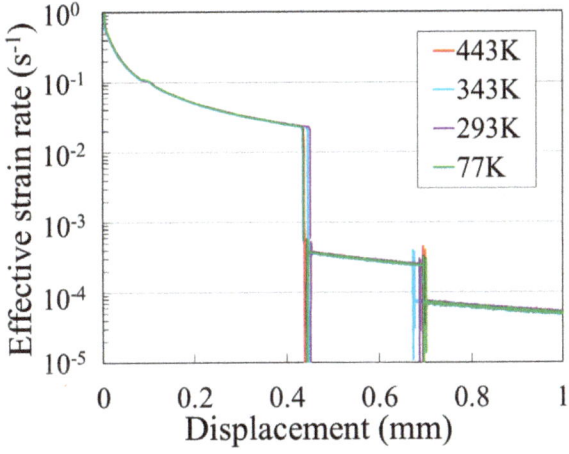

Figure 6. Effective strain rate–displacement relationship at each testing temperature.

We calculated the loading curvature-displacement relationship from the load–displacement relationship using Equation (2) (see Figure 7). The effect of temperature and strain rate on the indentation was confirmed by the change in the loading curvature. When testing temperatures greater than the room temperature, the loading curvature was observed to increase as the temperature increased regardless of the change in the displacement rate. At 293 and 343 K, an increase in the loading curvature was observed after holding as compared with that before holding. By contrast, there was a decrease in the loading curvature after holding for the test conducted at 77 and 443 K. At 77 K, not only the time in which the displacement rate increases but also the loading curvature decreases with increasing displacement.

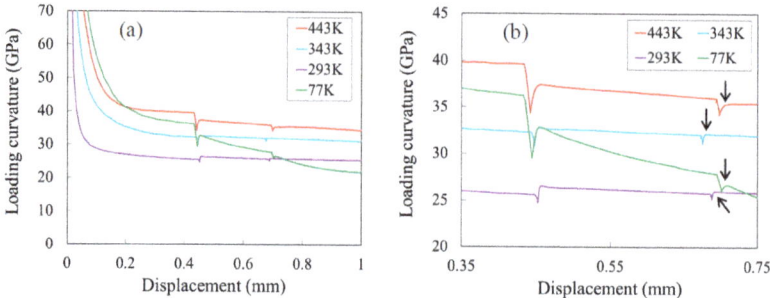

Figure 7. (a) Loading curvature-displacement curves at 77, 293, 343, and 443 K; (b) enlarged view. The change in displacement rate during indentation is as shown in Figure 4.

4. Discussion

4.1. Effect of Aging on the Material Strength

The precipitation process of 7075 alloys established in previous research is as follows: [32–34].

$$\text{Supersaturated solid solution (ssss)} \rightarrow \text{vacancy-rich clusters (VRC)} \rightarrow \text{GP zone} \rightarrow \eta' \rightarrow \eta. \quad (9)$$

A cluster or GP zone is an aggregate of atoms with a diameter of the order of nanometres. Herein, η' denotes the metastable phase, whereas η denotes the stable phase. Clusters and GP zones are formed during the natural and artificial aging, indicating an increase in material strength. After additional aging, η' precipitates and the strength of the material reaches its climax (peak aged). However, as η' continues to grow and η starts to precipitate, there is a decrease in the strength of the sample (over-aged). Therefore, increasing the testing temperature increases the strength as a result of the formation of precipitates until the peak-age condition is reached. The amount of solute atoms is also decreased. The succeeding sections discuss each testing temperature based on the above findings.

4.1.1. Temperature of 77 K

In [15], when the load was held during indentation at room temperature, the loading curvature after holding was obtained to be higher than that before the holding. This is attributed to the fact that an amount of solute atoms in solid solution diffuses into the dislocations during the holding period, thereby, causing the pinning of dislocations. At 77 K, which is a very low-temperature environment, the above-mentioned precipitation process was not observed and the sample probably remained in a solid solution. Therefore, solute atoms were in the solid solution during the indentation. However, there was no increase in the loading curvature after holding. This indicates that solute atoms have not segregated (or diffused) to the dislocations and pinned them in this low-temperature environment because solute atom diffusion is very slow as shown in Table 2, i.e., no time for diffusion.

4.1.2. Temperature of 293 and 343 K

In agreement with the results obtained in a previous study [18], an increase in the value of the loading curvature after holding was observed at 293 and 343 K as a result of dislocation pinning caused by the diffusion of solute atoms into the dislocations. At 343 K, the GP zone formed inside the material as a result of aging. However, a sufficient number of solute atoms to cause dislocation pinning was expected to be retained in the solution.

4.1.3. Temperature of 443 K

In contrast to the results obtained in the tests conducted at 293 and 343 K, at 443 K, the value of the loading curvature after holding was smaller than that before holding. This could be attributed to the fact that the amount of solute atoms must have been significantly reduced by aging, hence, the dislocation pinning effect was less likely to occur.

4.2. Effect of Testing Temperature on the Serration Behavior

Figure 8 shows an enlarged view of the region of the curve that is indicated by the arrow in Figure 7b as a means to investigate the details of the loading curvature-displacement relationship. Serrations were observed at 293 and 343 K but not at 77 and 443 K. The effective strain rate was approximately 7×10^{-4} s^{-1} at all testing temperatures (see Figure 6). It has been stated that the serrations observed in uniaxial tests were affected by the strain rate and the testing temperature. To discuss the effect of testing temperature on the serration behavior observed in this study, the testing temperatures related to indentation were investigated.

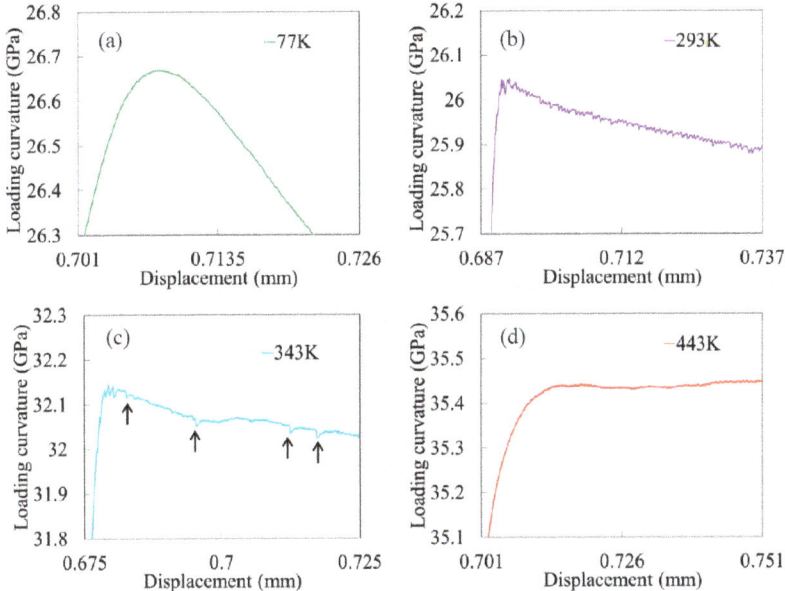

Figure 8. Enlarged view of Figure 7b: (**a**) 77 K, (**b**) 239 K, (**c**) 343 K, and (**d**) 443 K.

4.2.1. Temperature of 77 K

At 77 K, no fluctuation was observed in the loading curvature, as shown in Figure 8a, hence, serrations did not occur at this temperature. Because serrations occur at room temperature, it is assumed that the effect was not due to strain rate. This indicates that aging does not occur at very low temperatures. It also shows that there was barely any interaction between the dislocations and solute atoms.

4.2.2. Temperature of 293 K

As shown in Figure 8b, there was a significant fluctuation in the loading curvature. The occurrence of B-type serrations was confirmed (see Figure 1). There was a large number of solid solution atoms in the sample as 293 K corresponds to the early stage of aging. Consequently, there was a tendency for interaction between dislocations and solute atoms in solid solution to occur. Thus, the occurrence of a significant number of serrations was confirmed.

4.2.3. Temperature of 343 K

At 343 K, as shown by the arrow in Figure 8c, the interval at which the loading curvature dropped is greater than that at 293 K (see Figure 8b). Therefore, A- and B-type serrations (see Figure 1) were confirmed to have occurred. As the strain rate was constant, other causes of change in the serration phenomenon were observed. These are inferred to be the formation of GP zones, thus, a decrease in the number of solute atoms, and the increase in testing temperature. When the amount of solute atoms decreases, the chance of interaction with dislocations decreases, hence, serrations barely occur.

4.2.4. Temperature of 443 K

No serration was observed at 443 K because of the formation of η' and the increase in the strength of the material. The amount of solute atoms was, therefore, greatly reduced compared with other testing temperatures. Thus, the interaction between dislocations and solid solution atoms was less likely to occur. In addition, the diffusion rate of solute atoms at 443 K is higher than that at 343 K, hence, it is inferred that the remaining solid solution solute atoms were pinned to dislocations and deviates from the conditions at which serrations could occur. Therefore, it is believed that the interaction between dislocations and solid solution atoms does not appear in the loading curvature.

5. Conclusions

In this study, to clarify the effect of temperature change on the resulting serrations during indentation tests, we performed milli-indentations on an aluminum alloy (7075 alloy) at various temperatures. The serration phenomenon during indentation was varied by controlling the number of precipitated phases based on the effect of natural and artificial aging. This variation was as a result of the interaction between dislocations and the solid solution atoms observed under the different testing temperatures and strain rate on indentation, similar to that observed in previously reported uniaxial test results. Therefore, the serration phenomenon can be investigated via sharp indentation tests, which is considered valuable as a non-destructive testing technique for evaluating the dynamic strain aging of next-generation metals.

This study focused on the interaction between dislocations and the number of solute atoms based on the varying temperature and strain rate in indentation tests. Thus, the effect of the number of solute atoms and the testing temperature cannot be separated. Therefore, qualitative evaluations (e.g., microstructure evaluation by transmission electron microscopy or X-ray diffraction) to study the conditions that may separate these effects is recommended for future studies.

Author Contributions: H.Y. and T.K. conceived, designed, and performed the experiments; N.O. considered from research results; H.Y. and T.K. drafted this paper. All authors have read and agreed to the published version of the manuscript.

Funding: This research was funded by The Light Metal Educational Foundation, Inc., Japan.

Acknowledgments: The authors would like to thank Hikaru Ootani for assisting with the conduction of the indentation test.

Conflicts of Interest: The authors declare no conflict of interest.

References

1. Pink, E.; Grinberg, A. Stress drops in serrated flow curves of Al5Mg. *Acta Metall.* **1982**, *30*, 2153–2160. [CrossRef]
2. Pink, E.; Webernig, W.M. Precipitation during serrated flow in AlZn5Mg1. *Acta Metall.* **1987**, *35*, 127–132. [CrossRef]
3. Abbadi, M.; Hähner, P.; Zeghloul, A. On the characteristics of Portevin–Le Chatelier bands in aluminum alloy 5182 under stress-controlled and strain-controlled tensile testing. *Mater. Sci. Eng. A* **2002**, *337*, 194–201. [CrossRef]

4. Picu, R.C.; Vincze, G.; Ozturk, F.; Gracio, J.J.; Barlat, F.; Maniatty, A.M. Strain rate sensitivity of the commercial aluminum alloy AA5182-O. *Mater. Sci. Eng. A.* **2005**, *390*, 334–343. [CrossRef]
5. McCormick, P.G. A model for the Portevin-Le Chatelier effect in substitutional alloys. *Acta Metall.* **1972**, *20*, 351–354. [CrossRef]
6. Estrin, Y.; Kubin, L.P.; Aifantis, E.C. Introductory remarks to the viewpoint set on propagative plastic instabilities. *Scr. Metall. Mater.* **1993**, *29*, 1147–1150. [CrossRef]
7. Robinson, J.M.; Shaw, M.P. Microstructural and mechanical influences on dynamic strain aging phenomena. *Int. Mater. Rev.* **1994**, *39*, 113–122. [CrossRef]
8. Kami, T.; Yamada, H.; Ogasawara, N. Effect of strain rate on serrated load of indentation in Al-Mg alloy. *Trans. JSME* **2017**, *83*, 17–261.
9. Fischer-Cripps, A.C. *Nanoindentation*, 3rd ed.; Springer: New York, NY, USA, 2011.
10. Doerner, M.F.; Nix, W.D. A method for interpreting the data from depth-sensing indentation instruments. *J. Mater. Res.* **1986**, *1*, 601–609. [CrossRef]
11. Cheng, Y.T.; Cheng, C.M. Scaling approach to conical indentation in elastic-plastic solids with work hardening. *J. Appl. Phys.* **1998**, *84*, 1284–1291. [CrossRef]
12. Dao, M.; Chollacoop, N.; Van Vliet, K.J.; Venkatesh, T.A.; Suresh, S. Computational modeling of the forward and reverse problems in instrumented sharp indentation. *Acta Mater.* **2001**, *49*, 3899–3918. [CrossRef]
13. Ogasawara, N.; Chiba, N.; Zhao, M.; Chen, X. Measuring material plastic properties with optimized representative strain-based indentation technique. *J. Solid Mech. Mater. Eng.* **2007**, *1*, 895–906. [CrossRef]
14. Kami, T.; Yamada, H.; Ogasawara, N.; Chen, X. Strain rate behavior of pure aluminum in conical indentation with different indenter control methods. *Int. J. Comput. Methods Exp. Meas.* **2017**, *6*, 515–526. [CrossRef]
15. Wei, B.C.; Zhang, L.C.; Zhang, T.H.; Xing, D.M.; Das, J.; Eckert, J. Strain rate dependence of plastic flow in Ce-based bulk metallic glass during nanoindentation. *J. Mater. Res.* **2007**, *22*, 258–263. [CrossRef]
16. Su, C.; Herbert, E.G.; Sohn, S.; LaManna, J.A.; Oliver, W.C.; Pharr, G.M. Measurement of power-law creep parameters by instrumented indentation methods. *J. Mech. Phys. Solids.* **2013**, *61*, 517–536. [CrossRef]
17. Sudharshan Phani, P.; Oliver, W. Ultra high strain rate nanoindentation testing. *Materials* **2017**, *10*, 663. [CrossRef]
18. Kami, T.; Yamada, H.; Ogasawara, N. Critical strain of the sharp indentation through serration behavior with strain rate. *Inter. J. Mech. Sci.* **2019**, *152*, 512–523. [CrossRef]
19. Liu, L.; Ogasawara, N.; Chiba, N.; Chen, X. Can indentation technique measure unique elastoplastic properties? *J. Mater. Res.* **2008**, *24*, 784–800. [CrossRef]
20. Thevenet, D.; Mliha-Touati, M.; Zeghloul, A. The effect of precipitation on the Portevin-Le Chatelier effect in an Al–Zn–Mg–Cu alloy. *Mater. Sci. Eng. A.* **1999**, *266*, 175–182. [CrossRef]
21. Xue, Z.; Huang, Y.; Hwang, K.C.; Li, M. The influence of indenter tip radius on the micro-indentation hardness. *J. Eng. Mater. Tech.* **2002**, *124*, 371–379. [CrossRef]
22. Ferragut, R.; Somoza, A.; Tolley, A.; Torriani, I. Precipitation kinetics in Al-Zn-Mg commercial alloys. *J. Mater. Process Technol.* **2003**, *141*, 35–40. [CrossRef]
23. Riley, D.M.; McCormick, P.G. The effect of precipitation hardening on the Portevin–Le Chatelier effect in an Al-Mg-Si alloy. *Acta Metall.* **1977**, *25*, 181–185. [CrossRef]
24. Saha, G.G.; McCormick, P.G.; Rama Rao, P. Portevin–Le Chatelier effect in an Al–Mn Alloy I: Serration characteristics. *Mater. Sci. Eng.* **1984**, *62*, 187–196. [CrossRef]
25. Ikeno, S.; Watanabe, T.; Tada, S. On the serration in Al-Mg alloys at elevated temperatures. *J. Jpn. Inst. Met.* **1983**, *47*, 231–236. [CrossRef]
26. Nakayama, Y. Effects of Mg concentration, test temperature and strain rate on serration of Al-Mg system alloys and cause of its generation. *J. Jpn. Inst. Met.* **2000**, *64*, 1257–1262. [CrossRef]
27. Fujikawa, S.; Hirano, K. Diffusion of ^{65}Zn in aluminum and Al–Zn–Mg alloy over a wide range of temperature. *Trans. JIM* **1976**, *17*, 809–818. [CrossRef]
28. Fujikawa, S.; Hirano, K. Diffusion of ^{28}Mg in aluminum. *Mater. Sci. Eng.* **1977**, *27*, 25–33. [CrossRef]
29. Anand, M.S.; Murarka, S.P.; Agarwala, R.P. Diffusion of copper in nickel and aluminum. *J. Appl. Phys.* **1965**, *36*, 3860–3862. [CrossRef]
30. Park, W.S.; Chun, M.S.; Han, M.S.; Kim, M.H.; Lee, J.M. Comparative study on mechanical behavior of low temperature application materials for ships and offshore structures: Part I—Experimental investigations. *Mater. Sci. Eng. A* **2011**, *528*, 5790–5803. [CrossRef]

31. Wang, Y.; Jiang, Z. Dynamic compressive behavior of selected aluminum alloy at low temperature. *Mater. Sci. Eng. A* **2012**, *553*, 176–180. [CrossRef]
32. Sha, G.; Cerezo, A. Early-stage precipitation in Al–Zn–Mg–Cu alloy (7050). *Acta Mater.* **2004**, *52*, 4503–4516. [CrossRef]
33. Buha, J.; Lumley, R.N.; Crosky, A.G. Secondary ageing in an aluminium alloy 7050. *Mater. Sci. Eng. A.* **2008**, *492*, 1–10. [CrossRef]
34. Cao, C.; Zhang, D.; Zhuang, L.; Zhang, J. Improved age-hardening response and altered precipitation behavior of Al-5.2Mg-0.45Cu-2.0Zn (wt%) alloy with pre-aging treatment. *J. Alloys Compd.* **2017**, *691*, 40–43. [CrossRef]

© 2020 by the authors. Licensee MDPI, Basel, Switzerland. This article is an open access article distributed under the terms and conditions of the Creative Commons Attribution (CC BY) license (http://creativecommons.org/licenses/by/4.0/).

Article

Quantifying the Contribution of Crystallographic Texture and Grain Morphology on the Elastic and Plastic Anisotropy of bcc Steel

Martin Diehl [1,*,†], **Jörn Niehuesbernd** [2] **and Enrico Bruder** [2]

1. Max-Planck-Institut für Eisenforschung GmbH, Max-Planck-Straße 1, 40237 Düsseldorf, Germany
2. Division of Physical Metallurgy, Materials Science Department, TU Darmstadt, Alarich-Weiss-Straße 2, 64287 Darmstadt, Germany; joern.niehuesbernd@gmx.de (J.N.); e.bruder@phm.tu-darmstadt.de (E.B.)
* Correspondence: m.diehl@mpie.de; Tel.: +49-211-6792-187
† Current address: Department of Materials Science and Engineering, University of California Los Angeles, Los Angeles, CA 90095, USA.

Received: 24 September 2019; Accepted: 25 October 2019; Published: 22 November 2019

Abstract: The influence of grain shape and crystallographic orientation on the global and local elastic and plastic behaviour of strongly textured materials is investigated with the help of full-field simulations based on texture data from electron backscatter diffraction (EBSD) measurements. To this end, eight different microstructures are generated from experimental data of a high-strength low-alloy (HSLA) steel processed by linear flow splitting. It is shown that the most significant factor on the global elastic stress–strain response (i.e., YOUNG's modulus) is the crystallographic texture. Therefore, simple texture-based models and an analytic expression based on the geometric mean to determine the orientation dependent YOUNG's modulus are able to give accurate predictions. In contrast, with regards to the plastic anisotropy (i.e., yield stress), simple analytic approaches based on the calculation of the TAYLOR factor, yield different results than full-field microstructure simulations. Moreover, in the case of full-field models, the selected microstructure representation influences the outcome of the simulations. In addition, the full-field simulations, allow to investigate the micro-mechanical fields, which are not readily available from the analytic expressions. As the stress–strain partitioning visible from these fields is the underlying reason for the observed macroscopic behaviour, studying them makes it possible to evaluate the microstructure representations with respect to their capabilities of reproducing experimental results.

Keywords: anisotropy; linear flow splitting; crystal plasticity; DAMASK; texture; EBSD

1. Introduction

The plastic deformation induced during processing of metallic materials typically results in strong crystallographic textures and, thereby, macroscopically anisotropic mechanical properties. The prime example for such a process is the cold rolling of metal sheets, which is used for processing such diverse materials as aluminum, magnesium and steel. As the anisotropic elastic and plastic behaviour induced by texture and grain morphology has a significant influence on formability and dimensional accuracy, it is imperative to account for the anisotropy when conducting high-precision metal forming simulations [1]. However, the direct multi-scale inclusion of all microstructure details is usually computationally prohibitive. Approaches to reduce the computational efforts include model order reduction schemes [2,3], the use of Statistically Similar Representative Volume Elements (SSRVEs) [4] and homogenization methods like the Relaxed Grain Cluster (RGC) scheme by Tjahjanto et al. [5]. Despite these efforts to include microstructure details, usually analytic yield surface descriptions are employed to include plastic anisotropy. The superior execution speed of analytic yield descriptions, though, comes often at the price of significant experimental efforts associated with calibrating their constitutive parameters. Especially for complex yield surface descriptions (Banabic [6] gives a detailed overview), which require to probe the materials response in multiple deformation modes, it is therefore desirable to (partly) replace experiments by micro-mechanical simulations using numerical methods such as the Finite Element Method (FEM) [7] or Fast FOURIER Transform (FFT) based spectral methods [8]. Gawad et al. [9] extended this concept in their "Hierarchical Multi-Scale Model" (HMS) by performing on-the-fly yield surface computations.

All approaches that aim at improving the quality of component-scale simulations by taking the average material response from the homogenized polycrystal response are based on two ingredients: a description of grain morphology and texture as well as a suitable model for the single crystal behaviour. In this study, different approaches for microstructure and texture representation (i.e., the first ingredient) are compared with respect to their ability to correctly predict the elastic and plastic anisotropy of a strongly textured material. The materials constitutive behaviour (i.e., the second ingredient) is described with a crystal plasticity formulation that is classified as "phenomenological" by Roters et al. [10].

It should be noted that CPFEM studies with similar aims have already been performed 20 to 30 years ago [11–14]. While most of these studies were focused on the investigation of texture development, the ability of the CPFEM approach to predict average mechanical properties was also shown. However, the increase in computational capacities allows now to redo such investigations with much higher spatial resolution. In this study, more than two million discrete points—each with its directly measured orientation—are used for each individual simulation while in the early days of CPFEM modeling even a few hundreds of thousands elements was associated with long computation times.

The material investigated here is a high-strength low-alloy (HSLA) steel processed by linear flow splitting. The linear flow splitting process, presented in detail by Groche et al. [15], is used to produce bifurcated profiles in an integral style. It enables the manufacturing of sheet metal products with improved quality at lower costs [16] in comparison to conventional, multistep production routes. From previous investigations by Bruder et al. [17] it is known that the microstructure of the produced profile has a crystallographic texture and grain morphology that resembles that of cold rolled body-centred cubic (bcc) steels [18]. With regards to further processing of parts produced by the novel linear flow splitting technique, an accurate description of the resulting anisotropic material properties and their implementation in metal forming simulations is of great importance for exploiting the full potential of this technique. In a previous study by Niehuesbernd et al. [19], the elastic anisotropy induced by linear flow splitting in the investigated HSLA steel has been characterized experimentally and compared to predictions from analytic models based on the measured crystallographic texture. It was shown that the effective orientation dependent YOUNG's modulus can be accurately predicted from the crystallographic information when the geometric mean is used to calculate the polycrystalline average from the the single crystal stiffness/compliance tensor. The values obtained from the geometric mean lie well in-between the upper bound resulting from the assumption of spatially constant strains introduced by Voigt [20] and the lower bound based on the assumption of spatially constant stress by Reuss [21]. In addition—and in contrast to other approaches such as the Hill [22] average—this averaging scheme gives the same results regardless whether the stiffness or the compliance tensor is used. However, the complete omission of the grain morphology might render this approach invalid for the elongated grains of the probed material (compare the work of Jöchen et al. [23]). The present study therefore aims at evaluating the impact of the grain aspect ratio on the elastic and plastic behaviour of strongly textured microstructures by means of full-field simulations. To this end, results from numerical simulations employing microstructure representations of different degrees of sophistication are compared to simple analytic, texture-based models.

The study is structured as follows—First, details of the investigated material, including production steps and employed characterization methods, are given. The following section deals with the used numerical simulation method and the employed approaches for constructing microstructure representations from experimental data. The results are presented in Section 4 and compared and discussed with respect to the performance of the various simulation approaches in Section 5. After that, the conclusions that can be drawn from the results and the associated discussion are presented. The study finishes with an outlook on how to improve the predictive quality of crystal plasticity simulations.

2. Material: Composition, Processing and Characterization

The investigated material is an H480LA HSLA steel with a carbon content of 0.07 wt.%; details of the material are presented by Niehuesbernd et al. [19]. The microstructure of the material in as-received condition consists of ferrite grains and small cementite particles at the grain boundaries. Linear flow splitting was carried out continuously in 10 stages to produce double-Y-profiles with 12 mm long and 1 mm thick flanges (see Figure 1) from the initial sheet with a thickness of 2 mm.

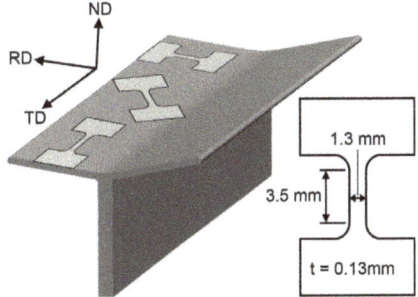

Figure 1. Upper half of the double-Y-profile produced by linear flow splitting with marked positions of the tensile samples (**left**) and their geometry (**right**).

Three mutually perpendicular cross sections parallel to normal direction (ND), rolling direction (RD) and transverse direction (TD) of the flanges were produced (see Figure 1) for texture and microstructure investigations. Sample preparation for Electron Backscatter Diffraction (EBSD) was performed using standard metallographic grinding and polishing techniques followed by an additional polishing step with an aqueous suspension of 0.05 µm Al_2O_3 particles. Subsequent EBSD measurements were carried out on all three samples with a Tescan Mira3 feg scanning electron microscope at a distance of 170 µm from the flange top surface. The size of the characterized area was adapted to the microstructure so that the maps contained at least 2000 grains and about 2.5 million measurement points to ensure an accurate representation of texture data and grain morphologies at the same time. The three obtained microstructure maps are shown in Figure 2 with color code assigned according to the inverse pole figure (IPF) in the respective sample surface normal direction. It can be seen that the material exhibits a microstructure with highly elongated, "pancake shaped" grains (see Figure 2) with average grain dimensions of 0.2 µm in ND, 0.8 µm in TD and 1.4 µm in RD. The apparent grain aspect ratios in the cross sectional measurements are therefore about 6.9 in the RD-section, 4.0 in the TD-section and 1.7 in the ND-section. The microstructure features a strong bcc-rolling texture including a distinct α-fiber ($\langle 110 \rangle$ || RD) with a dominant rotated cube orientation ($\{001\}\langle 110\rangle$ (the $\{001\}$ crystal planes are parallel to the sheet plane (ND) and the $\langle 110\rangle$ crystallographic directions are parallel to the rolling direction (RD).) having maximum intensity of about 20 times random and a typical γ-fiber ($\langle 111\rangle$ || ND). The $\varphi_2 = 45°$-section of the orientation distribution function (ODF) of the texture data from the TD-section is shown in Figure 3.

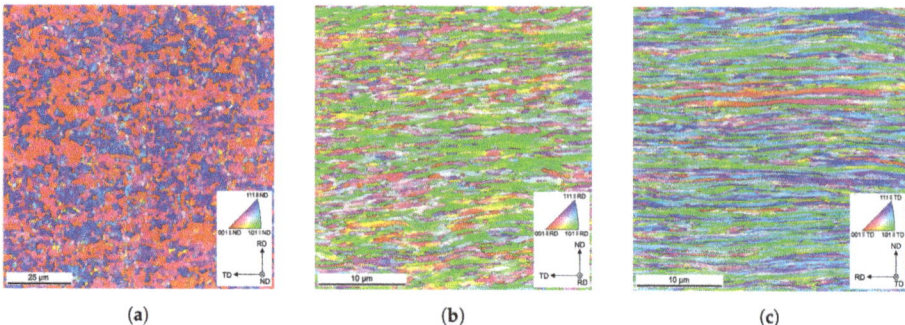

Figure 2. Microstructure maps in three mutually perpendicular directions of the material after linear flow splitting. Crystallographic orientation is given in terms of the inverse pole figure parallel to the measurement direction. Note the lower magnification of the normal direction (ND)-section in comparison to the rolling direction (RD)- and transverse direction (TD)-section. (**a**) ND-section. (**b**) RD-section. (**c**) TD-section.

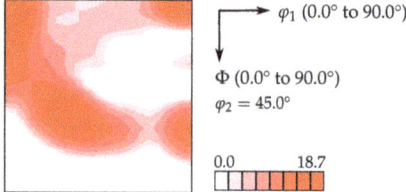

Figure 3. φ_2-section of the orientation distribution function (ODF) calculated from the TD-section using a harmonic series expansion approach. φ_1, Φ and φ_2 are the BUNGE–EULER angles.

Tensile tests were performed on the flange material in order to obtain experimental data on the plastic behaviour. For this purpose, dogbone-shaped tensile samples along TD, RD and under 45° between these directions were prepared (see Figure 1). The samples were ground from the flange top surface by 90 µm and afterwards from the lower surface to a final thickness of 130 µm in order to perform the tests at approximately the same positions as the microstructure investigations.

Without using numerical simulations, the orientation dependent YOUNG's modulus was directly estimated from the measured texture by computing the geometric mean of the stiffness tensor as:

$$\mathbf{C}_{\text{geom}} = \exp\left(\frac{1}{N}\sum_{i=1}^{N} \ln\left(\mathbf{T}_i^T \mathbf{C}\, \mathbf{T}_i\right)\right). \tag{1}$$

Here, N is the number of measurement points, $\mathbf{C}$ the stiffness tensor in crystal coordinates (cube orientation) and $\mathbf{T}$ are rotation matrices obtained from the EBSD measurements. The YOUNG's modulus in any given direction is then calculated from this tensor for each direction. As shown by Niehuesbernd et al. [19], values provided by this approach fall well into the range determined from ultrasonic measurements, which is therefore preferred over more involved approaches [22,24].

Given the success of this averaging approach when calculating the elastic response, it was also used to calculate the average Taylor [25] factor M for prediction of the average plastic behaviour. To this end, the individual TAYLOR factor M_i for uniaxial tension in the considered loading direction was calculated assuming slip on $\langle 1\,1\,1\rangle\{1\,1\,0\}$ and $\langle 1\,1\,1\rangle\{1\,1\,2\}$ slip systems with equal critical shear stresses on all slip systems for all orientations. Then, the geometric mean of these N TAYLOR factors is calculated according to the following equation:

$$M_{geom} = \exp\left(\frac{1}{N}\sum_{i=1}^{N} \ln(M_i)\right). \qquad (2)$$

The proof stress at 0.05% plastic deformation, σ_y, from the tensile test along TD was selected to determine the apparent critical resolved shear stress τ_{CRSS}. With the TAYLOR factor from the combined texture data of all three EBSD measurements a value of $\tau_{CRSS} = 268$ MPa was determined via $\tau_{CRSS} = \sigma_y/M$. This calculation is, however, only a rather rough approximation since it is based on the assumption of a homogeneous deformation of all points, irrespective of their crystallographic orientation. Moreover, this approximation does not take into account that different types of slip systems can have different critical resolved shear stresses. Nevertheless, this approach enables to analytically estimate the yield strength distribution for comparison with values obtained by numerical simulations and tensile tests.

3. Simulation Setup

The simulation setup, consisting of a microstructure representation, a constitutive law and a numerical solver for solving mechanical equilibrium under given boundary conditions, is outlined in the following.

3.1. Microstructure Representation

To investigate the influence of grain morphology and crystallographic texture on the global and local stress–strain behaviour, different microstructure representations are created based on the EBSD measurements presented in Section 2. While the first series of representations (I) is based on the individual data per measurement, all three measurements are combined for the second series (II) to increase the statistical reliability.

The five microstructure representations of series I based on the the three individual measurements are the following:

I a **Direct takeover 2D:** These 2D full-field models are based on a direct takeover of the measured crystallographic orientation on each of the 1601 × 1600 = 2,561,600 points (see Figure 2).
I b **Random orientation assignment 2D:** By randomly shuffling the measured crystallographic orientations among the points, a second set of 1601 × 1600 resolved 2D microstructures has been created.
I c **Random orientation assignment 3D:** The random distribution of almost all (Less than 2% of the discrete crystallographic orientations had to be discarded when distributing them on an equi-gridded cube ($136^3 < 1601 \times 1600 < 137^3$).) measured orientations on a 3D grid with 136 × 136 × 136 = 2,515,456 points gives a third set of microstructure variants.

The latter two microstructure variants lack any information on grain morphology but contain the full information of the crystallographic texture. This can be clearly seen in Figure 4a, where the 3D model (I c) based on the ND-section data is shown. When applied to a component scale simulation, this approach results in microstructure representations similar to the ones used in the "Texture-Component Crystal Plasticity FEM" (TCCP-FEM) introduced by Roters and Zhao [26] and Böhlke et al. [27].

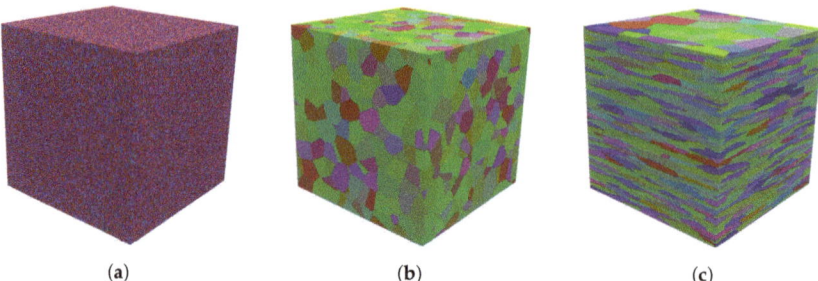

Figure 4. Microstructural models created from the measured crystallographic orientation. ND is aligned with the vertical direction, morphologically there is no difference between RD and TD for all three models. (**a**) Microstructure I c: Point-wise random orientation distribution, exemplarily shown for the ND-section. The legend is shown in Figure 2a. (**b**) Microstructure I e: 1000 globular grains with homogeneous crystallographic orientation, exemplarily shown for the RD-section. The legend is shown in Figure 2b. (**c**) Microstructure II c: 1000 elongated grains with homogeneous crystallographic orientation. The legend is shown in Figure 2c.

The orientation information, that is, texture, for the fourth and fifth set of microstructure representation is created in the following way: First, a discrete ODF with a bin size of 5.0° is created from the BUNGE–EULER angle representation of the crystallographic orientations without taking the sample symmetry into account. Second, using the HYBRIDIA method developed by Eisenlohr and Roters [28], the 1000 orientations that best represent the whole ODF are selected (see Reference [29] for a different approach to reduce the orientation data.). A comparison of texture index and entropy using MTEX 4.5.0 by Bachmann et al. [30] between the full texture and the selected orientation reveals a good approximation, especially there is no significant sharpening or weakening of the texture when using the approximation by 1000 orientations. This reduced texture is used for the following two representations in the first series:

I d **2D VORONOI tessellation:** A regular grid of 2024 × 2024 = 4,096,576 pixel is divided into 1000 grains with a periodic VORONOI tessellation. Each grain gets a homogeneous initial orientation assigned.

I e **3D VORONOI tessellation:** Similarly, a 160 × 160× 160 = 4,096,000 voxel grid is divided into 1000 equiaxed grains with a periodic VORONOI tessellation. The resulting microstructure for the RD-section is shown in Figure 4b.

Three more microstructure representations are generated from the combined texture information of all three measurements to increase the statistical reliability. The same approach to reduce the texture data to 1000 orientations as for microstructures I d and I e is employed:

II a **3D microstructure without grain information:** This TCCP-FEM model is conceptually a combination of variant I c (Random orientation assignment 3D) and I e (3D VORONOI tessellation): 1000 orientations are assigned to the points of a 10 × 10 × 10 grid.

II b **3D microstructure with globular grains:** The same geometric representation as for variant I e (3D VORONOI tessellation) is used but the 1000 orientations represent the texture of all three measurements. To investigate the influence of the grain shape separately from the influence of the strong crystallographic texture present in the probed material, a variant of this microstructure is created in which 1000 randomly sampled orientations are assigned to the grains.

II c **3D microstructure with elongated grains:** To generate elongated grains, a standard VORONOI tesselation of 1000 seed points is performed on a 160 × 160 × (160 · 8) grid from which only every eights plane along the last direction is used. The resulting grain structure with a grain aspect ratio of 8:8:1 (RD:TD:ND) and initial homogeneous orientation per grain is shown in Figure 4c.

To investigate the influence of the grain shape separately from the influence of the strong crystallographic texture present in the probed material, a variant of this microstructure is created in which 1000 randomly sampled orientations are assigned to the grains.

Preliminary control simulations have shown that the artificially created microstructures (I b to I e and II a to II c) are representative, that is, the statistical and macroscopic results considered here do not differ significantly. This finding is in agreement with a similar study on Dual Phase (DP) steels by Diehl [31] where measured microstructures where systematically coarsened.

3.2. Constitutive Model for Crystal Plasticity

A viscoplastic phenomenological formulation for crystal plasticity, introduced in similar form by Hutchinson [32] and Peirce et al. [33], is used in combination with an elastic stiffness tensor with cubic symmetry to describe the behaviour of the bcc material. This crystal plasticity model is based on the assumption that plastic slip γ occurs on a slip system α when the resolved shear stress τ^α exceeds a critical value ξ^α. The critical shear stress on each of the 24 slip systems is assumed to evolve from an initial value, ξ_0 to a saturation value ξ_∞ due to slip on the 12 $\langle 111 \rangle \{110\}$ and 12 $\langle 111 \rangle \{112\}$ systems according to the relation $\dot{\xi}^\alpha = h_0 |\dot{\gamma}^\beta| |1 - \xi^\beta/\xi_\infty^\beta|^a \, \text{sgn}(1 - \xi^\beta/\xi_\infty^\beta) h_{\alpha\beta}$ with initial hardening h_0, interaction coefficients $h_{\alpha\beta}$, a numerical parameter a and $\beta = 1, \ldots, 24$. The shear rate on system α is then computed as $\dot{\gamma}^\alpha = \dot{\gamma}_0 |\tau^\alpha/\xi^\alpha|^n \, \text{sgn}(\tau^\alpha/\xi^\alpha)$ with the inverse shear rate sensitivity n and reference shear rate $\dot{\gamma}_0$. The sum of the shear rates on all systems determines the plastic velocity gradient $\mathbf{L}_p$ in the employed finite strain formulation. Values for the single crystal stiffness tensor of iron at room temperature are known with good precision from experiments [34,35]. Here the values from the latter reference, given in Table 1a, are used. Parameters for the plastic behaviour (Table 1b) are based on parameters used by Tasan et al. [36], however, ξ_0 and ξ_∞ have been re-scaled by a constant factor such that model II c (3D microstructure with elongated grains) loaded in TD-direction reproduces the experimentally obtained proof stress. The constitutive formulation is implemented in the Düsseldorf Advanced Material Simulation Kit (DAMASK, presented in detail by Roters et al. [37,38]) where it can be used with different solvers for mechanical equilibrium, i.e., the commercial finite element solvers MSC.Marc and Abaqus and an efficient FFT-based spectral solver. The latter one is used in this study, details are given in the following.

Table 1. Constitutive parameters for the phenomenological crystal plasticity description. (a) Elastic behaviour. (b) Plastic behaviour.

(a)

Property	Value	Unit
C_{11}	230	GPa
C_{12}	134	GPa
C_{44}	116	GPa

(b)

Property	Value	Unit
$\dot{\gamma}_0$	1.0	mms
$\tau_{0,\{110\}}$	354	MPa
$\tau_{\infty,\{110\}}$	837	MPa
$\tau_{0,\{112\}}$	361	MPa
$\tau_{\infty,\{112\}}$	1538	MPa
h_0	1.0	GPa
Coplanar $h_{\alpha\beta}$	1.0	
Non-coplanar $h_{\alpha\beta}$	1.4	
n	20.0	
a	2.0	

3.3. Numerical Solver and Boundary Conditions

An FFT-based spectral solver is employed to solve for static mechanical equilibrium. It is based on the finite strain extension by Lahellec et al. [39] of the well-established formulation by Moulinec and Suquet [40], Lebensohn [41]; details regarding formulation, implementation and numerical performance are presented in References [42,43]. This solver operates on a regular grid, which allows the direct point-wise takeover of the EBSD data. Since an infinite medium is assumed, the data is periodically repeated in all three directions, which introduces artifacts at the boundary if the investigated microstructure is not periodic. For an infinite body, the applied boundary conditions are volume averages which in the employed large-strain formulation are given in mutually exclusive components of deformation gradient **F** and first PIOLA–KIRCHHOFF stress **P**. Uniaxial loading along 16 different directions at a rate of $0.0002\,\text{s}^{-1}$ was applied in 25 increments of 1 s duration, i.e., until a final technical strain of 0.5% was reached. In case of loading the ND-section (Figure 2a), loading varied from $\theta = 0.0°$ (along RD, horizontal) to $\theta = 168.75°$ in 11.25° steps, i.e., $\theta = 90.0°$ corresponds to loading along TD (vertical direction) and a rotation by $\theta = 180.0°$ is equivalent to no rotation ($\theta = 0.0°$). The corresponding deformation gradient, first PIOLA–KIRCHHOFF stress tensor and rotation matrix read as

$$\mathbf{F} = \begin{pmatrix} 1.0 + x & 0.0 & 0.0 \\ 0.0 & * & 0.0 \\ 0.0 & 0.0 & * \end{pmatrix} \tag{3a}$$

$$\mathbf{P} = \begin{pmatrix} * & * & * \\ * & 0.0 & * \\ * & * & 0.0 \end{pmatrix} \tag{3b}$$

$$\mathbf{R} = \begin{pmatrix} +\cos(\theta) & -\sin(\theta) & 0.0 \\ +\sin(\theta) & +\cos(\theta) & 0.0 \\ 0.0 & 0.0 & 1.0 \end{pmatrix}. \tag{3c}$$

Here, the symbol "*" indicates an undefined component since values in **F** and **P** are mutually exclusive. The strain x in the (11) component of **F** is set to 0.005 (0.5%) and θ measures the angle between RD and TD along ND.

4. Results

The simulation results are presented in the following. First, to quantify the average elastic and plastic behaviour, the orientation dependent YOUNG's modulus (E) and yield stress (σ_y) are given and compared to the corresponding results from the analytic calculations (Section 4.1). Then the local stress–strain distribution of selected simulations is presented in Section 4.2 to investigate in detail the differences at the micro-scale caused by the very different model assumptions.

4.1. Average Behaviour

YOUNG's modulus E resulting from the simulations is calculated as $E = \sigma/\varepsilon$ where σ is the average second PIOLA–KIRCHHOFF stress and ε the average GREEN–LAGRANGE strain along the loading direction at the first, purely elastic loading step.

Table 2 gives an overview of the obtained values for loading along ND, RD and TD. Table 2a shows that the simulation results obtained from the individual sections differ by at most +4 GPa and −3 GPa from the analytic results and Table 2b reveals even slightly smaller differences when using the combined texture (+3 GPa and −2 GPa). For both, analytic calculation and simulated results, the YOUNG's modulus along ND calculated from the RD-section is approximately 10 GPa higher than the value obtained from the TD-section. The differences between these sections are, hence, significantly higher than among all full-field simulation approaches.

Table 2. YOUNG's modulus E along ND, RD and TD. Niehuesbernd et al. [19] determined $E_{ND} = (204 \pm 10)$ GPa, $E_{RD} = (212 \pm 10)$ GPa and $E_{TD} = (232 \pm 10)$ GPa by ultrasonic measurements. (**a**) Results from the geometric mean calculation using the texture of the individual measurements. The highest and lowest values from simulations I a (direct takeover 2D), I b (random orientation assignment 2D), I c (random orientation assignment 3D), I d (2D VORONOI tessellation) and I e (3D VORONOI tessellation) are given as superscript and subscript, respectively. (**b**) Results from the geometric mean calculation and from simulations using the combined texture information. II a: 3D microstructure without grain information, II b: 3D microstructure with globular grains, II c: 3D microstructure with elongated grains.

(a)

	ND-Section	RD-Section	TD-Section
E_{ND}/GPa	-	205^{206}_{202}	194^{195}_{191}
E_{RD}/GPa	217^{220}_{219}	-	215^{217}_{214}
E_{TD}/GPa	233^{237}_{234}	231^{235}_{231}	-

(b)

	Geometric Mean		Simulation		
	All Orientations	1000 Orientations	II a	II b	II c
E_{ND}/GPa	198	198	199	198	196
E_{RD}/GPa	215	215	216	216	215
E_{TD}/GPa	233	234	235	234	236

Figure 5 displays the course of YOUNG's modulus over the three mutually perpendicular sections corresponding to the measurements. As the symmetry of grain shape and crystallographic texture allows to average the values of loading directions with an angular difference of 90° around the sample normal, only values for half of the considered loading direction range (0° to 180°) are shown. A cubic spline interpolation was performed to obtain values between the rotation angles for which a simulation was conducted. The analytic calculation has been performed at steps of 1°, making an interpolation unnecessary. Figure 5a compares the results of the analytic calculation to both 2D simulations using the full set of orientations from the individual measurements (i.e., microstructure sets I a and I b). Additionally, the range observed among all five simulations (I a to I e) is given as a background color. Figure 5b shows results from the analytic and numerical calculations from the combination of the full texture information and the cases of a random texture (models II b and II c only).

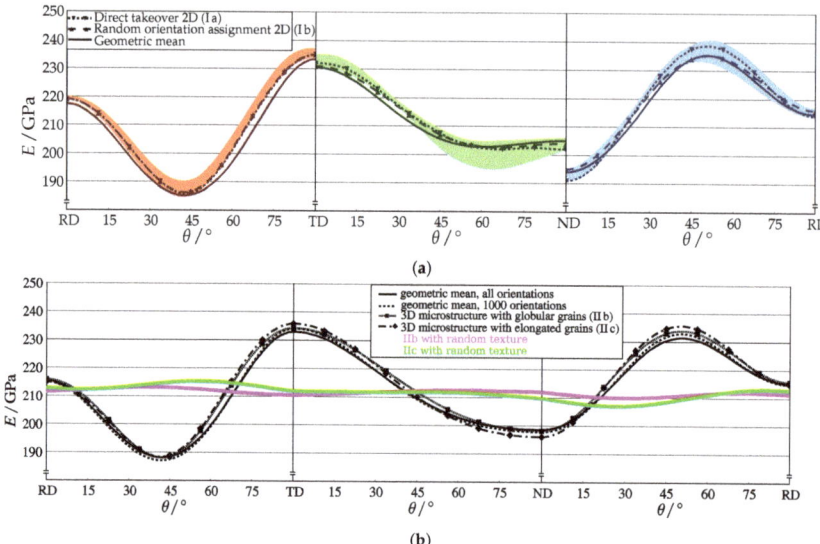

Figure 5. YOUNG's modulus in dependence of loading direction. Left: ND-section, Center: RD-section, Right: TD-section. (a) Results from simulations and the geometric mean calculation using the data of the individual measurements. The range between highest and lowest simulation result from all five microstructure variants (model I a to I e) is indicated by the background color (b) Results from the simulations and the geometric mean calculations using the combined texture data.

Among all simulation results obtained from the individual measurements (Figure 5a) the relative difference computed as $(\max(a_i) - \min(a_i))/\mathrm{avg}(\sum a_i)$ is smaller than 2.0%, 3.0% and 4.0% for the RD-section, ND-section and the TD-section, respectively. Results obtained by the analytic calculation are very close to the simulation not taking the grain shape into account (microstructure variant I b). The largest deviations between the two simulation approaches in Figure 5a can be seen for loading along ND (RD-section at 90°, TD-section at 0°), where the values obtained from the simulation including grain shape are lower by 4 GPa and at 45° between ND and RD where the simulation including grain shape is higher by 4 GPa. Overall, the influence of the grain morphology is rather small, a finding in agreement with a study by Jöchen et al. [23].

There are virtually no differences observable between the results from the analytic calculations using the complete orientation information obtained from all three measurements and its sample consisting of 1000 representative orientations, see Figure 5b. The same holds for the 3D models, where the use of globular (II b) and elongated (II c) grain shapes gives virtually the same results. Moreover, the differences between the simulations and the analytic calculations are smaller than 3 GPa (less than 1.5%) for the whole orientation range (Figure 5b).

The results from the tensile tests of the three samples from the flange material (Figure 1) are given in Figure 6; Figure 6a shows the engineering stress–strain curves and Figure 6b the extracted flow curves together with the 0.05% proof stress used to approximate the yield point. A clear influence of the loading direction can be seen: the sample oriented under 45° between RD and TD shows a significantly higher uniform elongation as well as a slightly higher strain hardening rate in comparison to the samples oriented in RD and TD, respectively. To determine the yield stress from the stress–strain curve, first the elastic portion of the strain is subtracted and the flow curves are plotted (Figure 6b). Then, the stress at 0.05% plastic strain was defined as the proof stress/yield point σ_y. The values determined in this manner amount to about 895 MPa in RD, 890 MPa in TD and 845 MPa under 45° rotation between RD and TD.

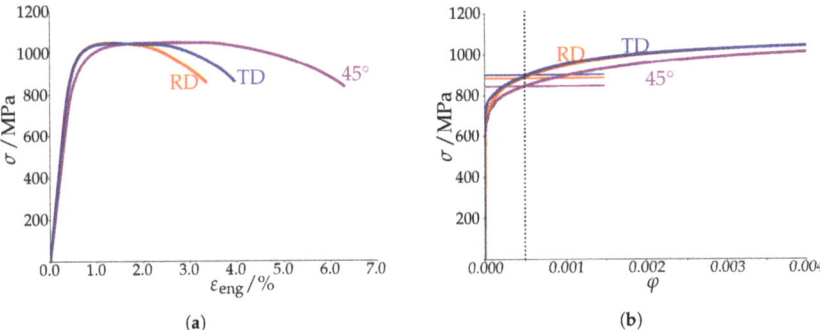

Figure 6. Experimental results of the tensile tests from the samples cut from the flange material. (a) Engineering stress–strain curves. (b) Flow curves the with 0.05% proof stress indicated. φ: Plastic deformation.

A similar but automated procedure was employed to define the yield stress of each of the 384 crystal plasticity simulations. For the automatic determination, first a continuous representation has been created with a spline interpolation from the 25 stress–strain values per simulation. From this smooth stress–strain curve, the elastic part has been subtracted to evaluate the stress at 0.05% plastic strain. A comparison with results obtained by the method proposed by Christensen [44] and the direct calculation of a plastic strain offset from the constitutive model (i.e., the plastic strain calculated from L_p) revealed only quantitative but no qualitative differences. It should be noted that adjusting the phenomenological constitutive parameters allows to reproduce the yield point or proof stress for any other method or threshold value as well.

Table 3 gives an overview of the obtained yield point values for loading along ND, RD and TD. In this table, the microstructure representations used to adjust the parameters are also indicated; those are the full orientation set for the TAYLOR factor calculation and variant II c for the full-field simulations. An influence of both, orientation data and modeling approach, is observed:

- The yield stress calculated for the individual sections with the analytic approach depends slightly on the data set, it differs by 30 MPa (i.e., 3.4%) for the yield stress in TD direction $\sigma_{y,TD}$, see Table 3a.
- The various microstructure models used for the individual data (I a to I e) predict differences of up to 38 MPa ($\sigma_{y,TD}$ calculated from ND-section data), see Table 3a.
- The yield stress in RD, $\sigma_{y,RD}$, predicted by all simulations is lower than the value obtained from the analytic expression.
- Sampling 1000 orientations from the combined texture results in an increase of the predicted yield stress by 4 MPa to 12 MPa when employing the analytic approach, see Table 3b.
- Employing the simpler models (II a: 3D microstructure without grain information and II b: 3D microstructure with globular grains) lowers $\sigma_{y,TD}$ and $\sigma_{y,ND}$ and increases $\sigma_{y,RD}$ in comparison to model II c (3D microstructure with elongated grains) which has the most realistic grain geometry, see Table 3b.

Table 3. Yield stress σ_y along ND, RD and TD. The experimental values are $\sigma_{y,RD} = 895$ MPa and $\sigma_{y,TD} = 890$ MPa. (**a**) Results from the geometric mean calculation using the texture of the individual measurements. The highest and lowest values from simulations I a (direct takeover 2D), I b (random orientation assignment 2D), I c (random orientation assignment 3D), I d (2D VORONOI tessellation) and I e (3D VORONOI tessellation) are given as superscript and subscript, respectively. (**b**) Results from the geometric mean calculation and simulations using the combined texture information. II a: 3D microstructure without grain information, II b: 3D microstructure with globular grains, II c: 3D microstructure with elongated grains. The results used to determine the scaling factor for the analytic expression and the crystal plasticity parameters from the experimental reference value are underlined.

(a)

	ND-Section	RD-Section	TD-Section
$\sigma_{y,ND}$/MPa	-	786_{876}^{902}	769_{832}^{862}
$\sigma_{y,RD}$/MPa	889_{844}^{873}	-	873_{818}^{837}
$\sigma_{y,TD}$/MPa	904_{873}^{911}	874_{885}^{898}	-

(b)

	Geometric Mean		Simulation		
	All Orientations	1000 Orientations	II a	II b	II c
$\sigma_{y,ND}$/MPa	778	782	857	853	877
$\sigma_{y,RD}$/MPa	874	883	842	837	825
$\sigma_{y,TD}$/MPa	<u>890</u>	902	888	885	<u>890</u>

The course of σ_y is presented in Figure 7 in a similar fashion as for the YOUNG's modulus in Figure 5. For σ_y, however, only results obtained from the combined texture data are presented as the inaccuracies resulting from the use of the individual measurements are already known. It can be seen that the two considered simulation approaches (II b and II c) form a narrow band (less than 15 MPa deviation) of yield point values and cross at four rotation angles. Although both analytic results are also close to each other, a clear difference to the simulation results can be seen. More precisely, the simulations predict a rather constant yield point from TD to ND (RD-section) and a peak between ND and RD whereas the TAYLOR factor calculation results in a decrease from TD to ND followed by a leveling-off increase between ND and RD. Qualitatively, the minimum at 45° between RD and TD is similarly predicted by the simulations and the analytic expression but the latter forecasts a higher value at RD. Comparison to the experimental results reveals a closer agreement for the crystal plasticity simulation at 45° between RD and TD and for the analytic expression at RD. When comparing the results of the simulations using a random texture, it can be seen that the grain morphology has only an effect when loading along ND, that is, perpendicular to the flat side of the elongated grains. More precisely, σ_y of the microstructure with elongated grains is higher by approximately 40 MPa in comparison to the microstructure having globular grains.

To investigate how the grain morphology influences the materials response at higher strain levels, the LANKFORD coefficient R, that is, the ratio between in-plane strain (perpendicular to the loading direction) and the out of plane strain (normal to the normal direction) is computed at a total strain of 10% in loading direction. Only 3D models using the combined and downsampled texture information are compared. The results are shown in Figure 8. It can be seen that the incorporation of the grain shape results in a reduced R value, while there is no significant difference whether spatially resolved globular grains or individual orientations per material point are used. It should be noted that the values of R depend critically on the used method, that is, which strain level is selected and whether the strain increments or the total strain is used for the determination.

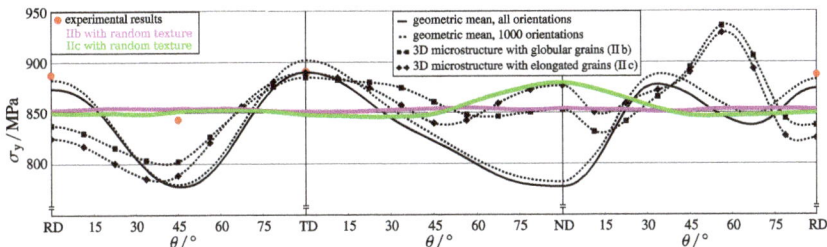

Figure 7. Yield point in dependence of loading direction. Left: ND-section, Center: RD-section, Right: TD-section. Results from the combined simulations obtained from the individual measurements and from the geometric mean calculations.

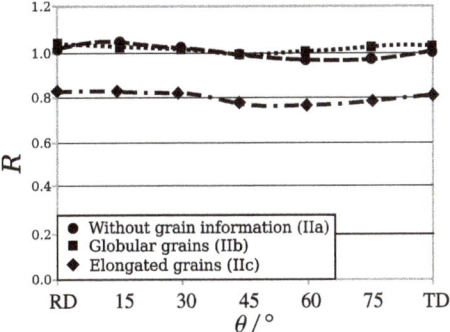

Figure 8. LANKFORD coefficient in dependence of the loading direction in the ND section. Results for the 3D models using the combined texture information (II a–c) are shown.

4.2. Micro-Mechanical Behaviour

The micro-mechanical behaviour presented in the following is based on the simulation results at step 20, that is, a strain of approximately 0.04%. This strain level corresponds to a stress just below the proof stress.

The spatial distribution of stress and strain in loading direction is shown exemplarily for the TD-section Figure 9, that is, a model of type I a. In this figure, the local stress and strain in loading direction is shown at 0.0°, 45.0°, 90.0° to ND. The grain structure is clearly visible, where the elongated grains are most obvious in the strain map when loaded perpendicular to the long grain axis and in the stress map when loaded along the long axis. A similar pattern can be observed for the RD-section (not shown in this study). The clear patterning ranging over the whole microstructure is less pronounced for equiaxed grains, that is, the ND section and VORONOI tessellated structures with globular grain morphology in two and three dimensions (I d, I e and II b). The pattern is totally missing for the random spatial distribution of crystallographic orientations (I b, I c and II a).

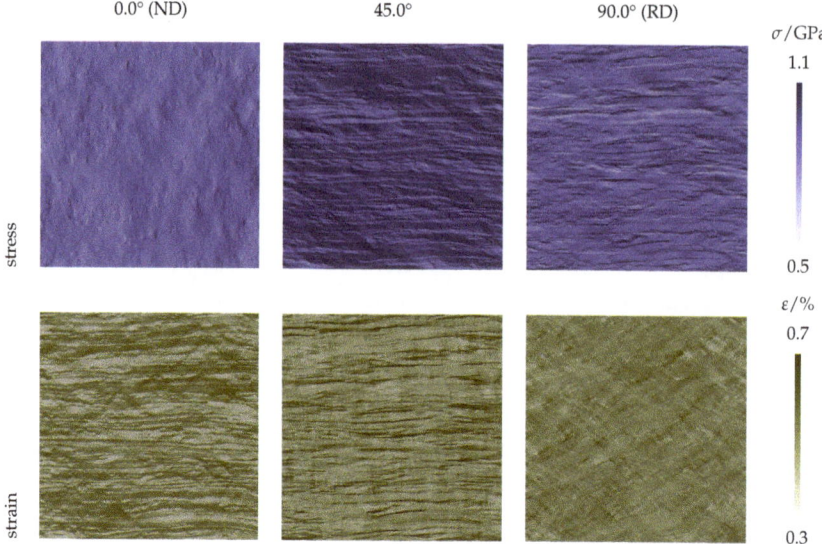

Figure 9. Stress (top row) and strain (bottom row) in loading direction for the TD-section (direct takeover, microstructure representation I a). The left image shows loading along ND (vertical direction), the right image loading along RD (horizontal direction) and the central loading aligned at $\Theta = 45°$ in-between ND and RD. A logarithmic mapping from value to color is employed for stress and strain.

For a more quantitative inspection that also enables to systematically investigate the 3D microstructures, "heat maps" of the stress–strain response of each voxel of the employed microstructures are plotted. In Figure 10, such maps are shown for the 2D microstructure models generated from all measured crystallographic orientations in the RD-section sample (model type I a and I b). Figure 10a,c show the stress–strain response for loading along TD, that is, along the elongated grains, for the model including grain morphology and the model with random distribution, respectively. The corresponding plots for loading along ND, that is, perpendicular to the long axis of the grains are given in Figure 10b,d. Independently of the microstructure model, a characteristic unimodal distribution results from the loading along TD while a bimodal distribution results from the loading along the ND. This bimodal distribution is approximately parallel to the strain axis and, hence, results in unimodal stress distributions (shown on the right side of the heat maps). In contrast, the shape of the strain distributions (shown on the top of the heat maps) depends on the microstructure model. Taking the grain shape into account (I a, Figure 10a) results in a bimodal distribution while the minimum deteriorates to a plateau for the random orientation assignment (I b, Figure 10d).

Figure 11 shows the heat maps for loading along ND from model variants I c (Random orientation assignment 3D), I d (2D VORONOI tessellation), I e (3D VORONOI tessellation) created from the RD-section and II c (i.e., using the combined texture information). Comparing Figure 11a with Figure 10d shows that for texture component modeling a difference in the stress–strain partitioning between the 2D and the 3D model (model I b and I c) is hard to ascertain. In contrast, using a 2D or a 3D model makes a difference for spatially resolved grains: The 2D variant of the model with 1000 grains (I d, Figure 11b) shows significantly stronger clustering than the 3D counterpart (I e, Figure 11c). The use of a realistic grain shape (II c, Figure 11d) narrows the strain distribution in comparison to the use of globular grains (I d, Figure 11c). The strain distribution is even more narrow when the measured microstructure is directly imported (I a, Figure 10b).

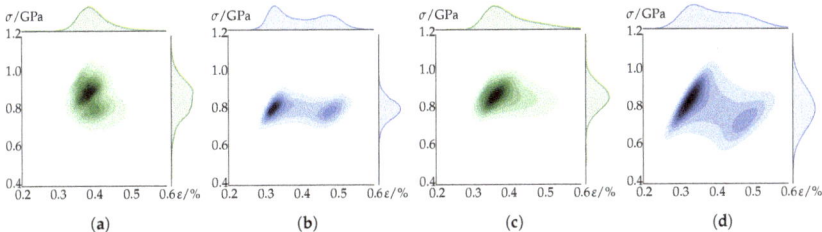

Figure 10. Distribution of the stress–strain correlation ("heat map") in models created from all crystallographic orientations measured in the RD-section for loading along TD and ND using a kernel density estimation. Note: Modeling the response by an isostrain assumption would result in a vertical line, the isostress assumption would result in an horizontal line. (**a**) Direct takeover 2D (I a), loading along TD. (**b**) Direct takeover 2D (I a), loading along ND. (**c**) Random orientation assignment 2D (I b), loading along TD. (**d**) Random orientation assignment 2D (I b), loading along ND.

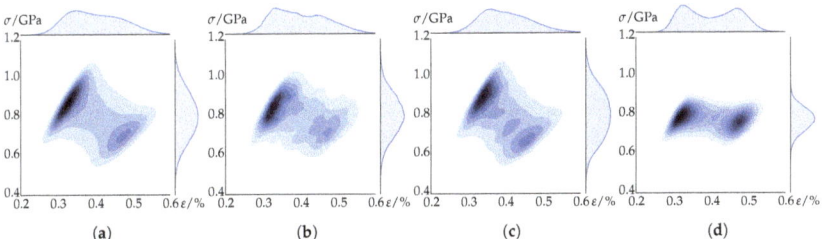

Figure 11. Distribution of the stress–strain correlation ("heat map") in models created from the measurement in the RD-section and in a model created from the combination of all three measurements for loading along ND using a kernel density estimation. Note: Modeling the response by an isostrain assumption would result in a vertical line, the isostress assumption would result in an horizontal line. (**a**) Random orientation assignment 3D (I c), created from the RD-section. (**b**) 2D VORONOI tessellation (I d), created from the RD-section. (**c**) 3D VORONOI tessellation (I e), created from the RD-section. (**d**) 3D model with elongated grains (II c).

5. Discussion

Based on the results presented in the previous section, the different approaches for predicting the global and local material response from experimental orientation data are discussed here. This discussion is based mainly on the local behaviour as it allows to quantify the factors influencing the internal stress and strain distribution which in turn determines the global response.

As revealed by the micro-mechanical investigations, the stress–strain distribution resulting from the full-field simulations has a characteristic shape for the different loading directions—a bimodal distribution results from loading along ND while unimodal distributions arise from loading along TD and RD. This behaviour is vastly independent of the selected microstructure representation, i.e., all presented numerical approaches result qualitatively in a similar distribution. It can, hence, be concluded that the materials response caused by crystallographic alignment with respect to the loading direction has a much stronger influence on the stress–strain response than the grain morphology. For the small strains (0.5%) and the rather isotropic plastic behaviour assumed in this study, the elastic constants are dominating the constitutive response. The local stress–strain behaviour can therefore be explained by the spread of the YOUNG's modulus in the respective direction. For loading in ND, most grains have either their $\langle 0\,0\,1 \rangle$ or $\langle 1\,1\,1 \rangle$ direction aligned with the loading direction. Since these crystals directions have vastly different directional YOUNG's moduli of 130 GPa and 275 GPa, respectively; the overall YOUNG's modulus distribution is characterized by two peaks. In contrast, for loading in

RD, most grains have either their $\langle 1\,1\,0\rangle$ or $\langle 1\,1\,2\rangle$ direction aligned with the loading direction. These crystallographic directions possess virtually the same directional YOUNG's modulus of around 210 GPa. Thus, in RD the overall YOUNG's modulus distribution shows only one narrow peak. The same holds true for loading along TD, however, the spread is slightly broader in this direction due to the presence of $\{1\,1\,2\}\langle 1\,1\,0\rangle$ orientations with a directional YOUNG's modulus of 275 GPa. As these characteristics of the elastic properties are fully taken into account when using the analytic expression to compute E. Therefore, this approach (Equation (1)) gives very accurate predictions and no improvement can be achieved by utilizing full-field methods that additionally consider the grain shape.

The deviations between the predictions from the different EBSD measurements have shown that the key factor for accurate predictions is the precise determination of the crystallographic texture. This usually requires probing a large volume of the material and computationally expensive simulations. However, the number of orientations required for the actual calculation can be drastically reduced by the use of an appropriate sampling strategy. Here it was shown that sampling 1000 orientations to approximate the 12,000,000 measurement points suffices to predict the YOUNG's modulus with an accuracy that exceeds the precision of ultrasonic measurements [19].

When predicting the plastic behaviour in terms of the yield stress, the choice of the microstructure model has a higher influence than in the elastic case. This can be attributed to the non-linear and rate-dependent plastic behaviour which is strongly influenced by the level and direction of plastic shear in the neighboring material points. These interactions are completely ignored when using the analytic expression based on the TAYLOR factor (Equation (2)). Hence, the observed deviations between this simple approach and the numerical predictions are to be expected. The largest deviation is seen for loading along ND, where the upper bound prediction of the analytic expression significantly exceeds the simulation results. The reason for this observation is the very inhomogeneous strain state which renders the underlying isostrain assumption of the analytic expression not suitable in this case. This holds especially for the prediction of the LANKFORD coefficient which is obtained at a significantly higher strain value. In contrast, the combination of a crystallographic texture with a unimodal distribution of elastic stiffness and a grain structure resembling an array of stacked disks, lead to an almost ideal isostress situation. Therefore, approaches that are not based on the isostrain assumption [45–47] or self-consistent approaches as introduced by Molinari et al. [48] and Lebensohn and Tomé [49] are expected to improve the prediction without relying on computationally costly full-field simulations. In that context, it should be mentioned that the increase of the yield strength for elongated grains in comparison to globular grains observed for the random texture (Figure 7) can not be explained by reasoning in terms of isostrain or isostress models. While the isostress model gives the lower bound for the elastic modulus, here a higher yield stress is observed for the more isostress-like situation of elongated grains. Analysis of the stress–strain data has shown that this is a result of the initially higher hardening rate of the microstructure with elongated grains which results in a higher proof stress for the (relatively large) offset of 0.05% strain.

Even though the full-field simulations are largely consistent among each other, the predicted yield stress for loading along RD is lower by 50 MPa in comparison to the experimental results. One possible reason for this discrepancy is the underlying assumption of a homogeneous initial material hardening state when setting up the simulation. To investigate how far this assumption is violated, the geometrically necessary dislocations (GNDs) in the ND-section microstructure were estimated using the approach of Field et al. [50]. Based on the median of the TAYLOR factor for the TD (reference direction for parameter adjustment) and the RD loading directions, the average GND density for low and high TAYLOR factors was computed. For loading along TD, no difference in the GND density between grains with low and high M values could be found. More precisely, the difference was less than 1%. In contrast, for loading along RD, the orientations with a lower M value showed an increase of the GND density by approximately 8% (As the GND density calculation from EBSD measurements is associated with multiple sources for errors, it is discussed here only in relative terms.). This indicates that the nominally "soft" grains for loading along RD are hardened more than their "hard" counterparts and, hence, the yield stress is underestimated in the simulations due to the assumption of a homogeneous initial material behaviour.

6. Conclusions

In the present study, the viability of simple analytical texture-based models is discussed and evaluated by comparing them with different numerical microstructure models and experimental data. As the model material used is a HSLA steel processed by linear flow splitting, special attention is paid to the characteristics of this material—namely the grain morphology and the cold rolling-type crystallographic texture—and their effect on the global and local stress–strain behaviour. The obtained results lead to the following conclusions:

- The grain morphology only has a minor impact on anisotropic elastic and plastic properties, with differences of less than 3% between microstructure based and solely texture based numerical models.
- Statistically sufficient orientation measurements are more important than grain morphology. Even measuring 2000 grains does not ensure obtaining a representative orientation data.
- The HYBRIDIA method enables a significant reduction of the orientation data that is required to accurately represent the texture.
- The simple analytic approach based on the geometric mean is suitable for estimating anisotropic elastic properties, since it yields very similar results as more complex numerical simulations.
- The underlying isostrain assumption of the TAYLOR model renders it an unsuitable choice for materials consisting of non-equiaxed grains with very strong anistropic behaviour.

These results indicate that full-field simulations are not required for predicting the YOUNG's modulus in dependence of the orientation. Even the simple averaging scheme used in this study predict values in agreement with experiments and full field simulations. Hence, more advanced averaging schemes Hill [22], Kiewel and Fritsche [24], Kiewel et al. [51] are unlikely to give better predictions. This finding holds also to a large extend for the plastic behaviour. Moreover, as full-field crystal plasticity simulations are often based on microstructures consisting of only a few hundred grains in an attempt to minimize the computational efforts, there is a danger of using "non-representative volume elements". Established mean-field homogenization approaches [52] such as the Grain Interaction Model (GIA) [53], the (A)LAMEL model [54], the Relaxed Grain Cluster (RGC) model [55,56] or self-consistent approaches [48,49] are, therefore, better suited as their computational performance does not require significant compromises on the number of crystallographic orientations. In many cases they also correctly predict the texture evolution after large deformation in good agreement with experimental results [49,54], a task that is especially challenging for full-field approaches due to the severe mesh deterioration. It should be noted, however, that the underlying assumptions of mean-field homogenization models make them less suited for materials with a high contrast in stiffness or strength, such as dual phase (DP) steels [57] or $\alpha + \beta$-titanium alloys [58].

7. Outlook

The presented findings allow also to draw conclusions for the further use of crystal plasticity simulations aiming at investigating the material response at the microstructure scale. As obvious from the investigated stress–strain correlations, the use of 2D microstructures results in more pronounced localization than in corresponding 3D microstructures and—as shown by [59,60]—makes any investigation of the local environment impossible. Given the fact that 3D characterizations, i.e., serial sectioning EBSD [61] or synchrotron measurements [62] are rather costly, the creation of artificial microstructures is a good compromise that takes both aspects, crystallographic orientation and grain morphology, into account. Decisive for such approaches is the approximation of the ODF with a rather small number of distinct orientations. The HYBRIDIA scheme by Eisenlohr and Roters [28] has shown a good performance for this task. The microtexture, however, was not taken into account in the present study. As preferential orientation relations between neighboring grains are present in most textured materials, considering the misorientation distribution function (MODF) introduced by Pospiech et al. [63] following the approach of Miodownik et al. [64] can further increase the similarity between real and synthetic microstructures. The same holds for the incorporation of in-grain orientation scatter. In the present study, the most realistic microstructure model considered only the average grain elongation. Even though this results already in a significantly improved local stress–strain response, taking the full grain size distributions into account would result in a significantly more realistic grain morphology. DREAM.3D, a software developed by Groeber et al. [65,66], Groeber and Jackson [67] provides tools for this purpose; the generated microstructures can be directly imported into DAMASK as shown by Diehl et al. [68]. Last but not least, the preexisting inhomogeneity of the hardening state among the different orientations/grains should be considered when setting up the simulation. While this is conceptually also possible with the employed phenomenological description, the use of a dislocation density-based model would allow to use directly the GND density from the EBSD measurements as an input parameter without additional fitting. The employed DAMASK package [38] offers a variety of such physics-based models for bcc materials [69], fcc steels [70] and Tungsten [71]. More advanced modeling approaches that allow to investigate in detail the influence of dislocation movement and interaction with grain boundaries [72] or the role of damage [73–75] are additionally available. However, the challenge remains to increase the computational performance of these models drastically before they can be employed to polycrystals that contain a sufficient number of grains to be statistically representative.

Author Contributions: Conceptualization, M.D., J.N., and E.B.; analytic calculations, J.N.; simulations, M.D.; experiments, J.N.; data curation, M.D.; writing—original draft preparation, M.D., J.N., and E.B.; writing—review and editing, M.D.; visualization, M.D.

Funding: J.N. and E.B. gratefully acknowledge the Deutsche Forschungsgemeinschaft (DFG) for founding the present work, which in part has been carried out within the Collaborative Research Centre 666 "Integral sheet metal design with higher order bifurcations—Development, Production, Evaluation".

Conflicts of Interest: The authors declare no conflict of interest.

References

1. Raabe, D.; Klose, P.; Engl, B.; Imlau, K.P.; Friedel, F.; Roters, F. Concepts for integrating plastic anisotropy into metal forming simulations. *Adv. Eng. Mater.* **2002**, *4*, 169–180. [CrossRef]
2. Fritzen, F.; Böhlke, T. Three-dimensional finite element implementation of the nonuniform transformation field analysis. *Int. J. Numer. Methods Eng.* **2010**, *84*, 803–829. [CrossRef]
3. Michel, J.C.; Suquet, P. A model-reduction approach to the micromechanical analysis of polycrystalline materials. *Comput. Mech.* **2016**, *57*, 483–508. [CrossRef]
4. Brands, D.; Balzani, D.; Scheunemann, L.; Schröder, J.; Richter, H.; Raabe, D. Computational modeling of dual-phase steels based on representative three-dimensional microstructures obtained from EBSD data. *Arch. Appl. Mech.* **2016**, *86*, 575–598. [CrossRef]
5. Tjahjanto, D.D.; Eisenlohr, P.; Roters, F. Multiscale deep drawing analysis of dual-phase steels using grain cluster-based RGC scheme. *Model. Simul. Mater. Sci. Eng.* **2015**, *23*, 045005. [CrossRef]
6. Banabic, D. *Sheet Metal Forming Processes*; Springer: Berlin/Heidelberg, Germany, 2010. [CrossRef]
7. Kraska, M.; Doig, M.; Tikhomirov, D.; Raabe, D.; Roters, F. Virtual material testing for stamping simulations based on polycrystal plasticity. *Comput. Mater. Sci.* **2009**, *46*, 383–392. [CrossRef]
8. Zhang, H.; Diehl, M.; Roters, F.; Raabe, D. A virtual laboratory for initial yield surface determination using high resolution crystal plasticity simulations. *Int. J. Plast.* **2016**, *80*, 111–138. [CrossRef]
9. Gawad, J.; Van Bael, A.; Eyckens, P.; Samaey, G.; Van Houtte, P.; Roose, D. Hierarchical multi-scale modeling of texture induced plastic anisotropy in sheet forming. *Comput. Mater. Sci.* **2013**, *66*, 65–83. [CrossRef]
10. Roters, F.; Eisenlohr, P.; Hantcherli, L.; Tjahjanto, D.D.; Bieler, T.R.; Raabe, D. Overview of constitutive laws, kinematics, homogenization, and multiscale methods in crystal plasticity finite element modeling: Theory, experiments, applications. *Acta Mater.* **2010**, *58*, 1152–1211. [CrossRef]

11. Marin, E.B.; Dawson, P.R. Elastoplastic finite element analyses of metal deformations using polycrystal constitutive models. *Comput. Methods Appl. Mech. Eng.* **1998**, *165*, 23–41. [CrossRef]
12. Beaudoin, A.J.; Dawson, P.R.; Mathur, K.K.; Kocks, U.F.; Korzekwa, D.A. Application of polycrystal plasticity to sheet forming. *Comput. Methods Appl. Mech. Eng.* **1994**, *117*, 49–70. [CrossRef]
13. Mathur, K.K.; Dawson, P.R. On modeling the development of crystallographic texture in bulk forming processes. *Int. J. Plast.* **1989**, *5*, 67–94. [CrossRef]
14. Barbe, F.; Decker, L.; Jeulin, D.; Cailletaud, G. Intergranular and intragranular behaviour of polycrystalline aggregates. Part 1: F.E. model. *Int. J. Plast.* **2001**, *17*, 513–536. [CrossRef]
15. Groche, P.; Vucic, D.; Jöckel, M. Basics of linear flow splitting. *J. Mater. Process. Technol.* **2007**, *183*, 249–255. [CrossRef]
16. Groche, P.; Bruder, E.; Gramlich, S. (Eds.) *Manufacturing Integrated Design*; Springer International Publishing: Cham, Switzerland, 2017. [CrossRef]
17. Bruder, E.; Ahmels, L.; Niehuesbernd, J.; Müller, C. Manufacturing-induced material properties of linear flow split and linear bend split profiles. *Mater. Werkst.* **2017**, *48*, 41–52. [CrossRef]
18. Hölscher, M.; Raabe, D.; Lücke, K. Rolling and recrystallization textures of bcc steels. *Steel Res. Int.* **1991**, *62*, 567–575. [CrossRef]
19. Niehuesbernd, J.; Müller, C.; Pantleon, W.; Bruder, E. Quantification of local and global elastic anisotropy in ultrafine grained gradient microstructures, produced by linear flow splitting. *Mater. Sci. Eng. A* **2013**, *560*, 273–277. [CrossRef]
20. Voigt, W. Über die Beziehung zwischen den beiden Elastizitätskonstanten isotroper Körper. *Ann. Phys.* **1889**, *38*, 573–587. [CrossRef]
21. Reuss, A. Berechnung der Fließgrenze von Mischkristallen auf Grund der Plastizitätsbedingung für Einkristalle. *Z. Angew. Math. Mech.* **1929**, *9*, 49–58. [CrossRef]
22. Hill, R. The Elastic Behaviour of a Crystalline Aggregate. *Proc. Phys. Soc. Sect. A* **1952**, *65*, 349. [CrossRef]
23. Jöchen, K.; Böhlke, T.; Fritzen, F. Influence of the Crystallographic and the Morphological Texture on the Elastic Properties of Fcc Crystal Aggregates. *Solid State Phenom.* **2010**, *160*, 83–86. [CrossRef]
24. Kiewel, H.; Fritsche, L. Calculation of effective elastic moduli of polycrystalline materials including nontextured samples and fiber textures. *Phys. Rev. B* **1994**, *50*, 5–16. [CrossRef] [PubMed]
25. Taylor, G.I. Plastic strain in metals. *J. Inst. Met.* **1938**, *62*, 307–324.
26. Roters, F.; Zhao, Z. Application of the Texture Component Crystal Plasticity Finite Element Method for Deep Drawing Simulations—A Comparison with Hill's Yield Criterion. *Adv. Eng. Mater.* **2002**, *4*, 221–223. [CrossRef]
27. Böhlke, T.; Risy, G.; Bertram, A. A texture component model for anisotropic polycrystal plasticity. *Comput. Mater. Sci.* **2005**, *32*, 284–293. [CrossRef]
28. Eisenlohr, P.; Roters, F. Selecting sets of discrete orientations for accurate texture reconstruction. *Comput. Mater. Sci.* **2008**, *42*, 670–678. [CrossRef]
29. Jöchen, K.; Böhlke, T. Representative reduction of crystallographic orientation data. *J. Appl. Crystallogr.* **2013**, *46*, 960–971. [CrossRef]
30. Bachmann, F.; Hielscher, R.; Schaeben, H. Texture Analysis with MTEX—Free and Open Source Software Toolbox. *Solid State Phenom.* **2010**, *160*, 63–68. [CrossRef]
31. Diehl, M. *High-Resolution Crystal Plasticity Simulations*; Apprimus Wissenschaftsverlag: Aachen, Germany, 2016.
32. Hutchinson, J.W. Bounds and self-consistent estimates for creep of polycrystalline materials. *Proc. R. Soc. A* **1976**, *348*, 101–127. [CrossRef]
33. Peirce, D.; Asaro, R.J.; Needleman, A. An analysis of nonuniform and localized deformation in ductile single crystals. *Acta Metall.* **1982**, *30*, 1087–1119. [CrossRef]
34. Rayne, J.A.; Chandrasekhar, B.S. Elastic Constants of Iron from 4.2 to 300 °K. *Phys. Rev.* **1961**, *122*, 1714–1716. [CrossRef]
35. Adams, J.J.; Agosta, D.S.; Leisure, R.G.; Ledbetter, H. Elastic constants of monocrystal iron from 3 to 500 K. *J. Appl. Phys.* **2006**, *100*, 113530. [CrossRef]
36. Tasan, C.C.; Hoefnagels, J.P.M.; Diehl, M.; Yan, D.; Roters, F.; Raabe, D. Strain localization and damage in dual phase steels investigated by coupled in-situ deformation experiments-crystal plasticity simulations. *Int. J. Plast.* **2014**, *63*, 198–210. [CrossRef]

37. Roters, F.; Eisenlohr, P.; Kords, C.; Tjahjanto, D.D.; Diehl, M.; Raabe, D. DAMASK: The Düsseldorf Advanced Material Simulation Kit for studying crystal plasticity using an FE based or a spectral numerical solver. In *Procedia IUTAM: IUTAM Symposium on Linking Scales in Computation: From Microstructure to Macroscale Properties*; Cazacu, O., Ed.; Elsevier: Amsterdam, The Netherlands, 2012; Volume 3, pp. 3–10. [CrossRef]
38. Roters, F.; Diehl, M.; Shanthraj, P.; Eisenlohr, P.; Reuber, C.; Wong, S.L.; Maiti, T.; Ebrahimi, A.; Hochrainer, T.; Fabritius, H.O.; et al. DAMASK—The Düsseldorf Advanced Material Simulation Kit for Modelling Multi-Physics Crystal Plasticity, Damage, and Thermal Phenomena from the Single Crystal up to the Component Scale. *Comput. Mater. Sci.* **2019**, *158*, 420–478. [CrossRef]
39. Lahellec, N.; Michel, J.C.; Moulinec, H.; Suquet, P. Analysis of Inhomogeneous Materials at Large Strains Using Fast Fourier Transforms. In *IUTAM Symposium on Computational Mechanics of Solid Materials at Large Strains*; Solid Mechanics and Its Applications; Miehe, C., Ed.; Kluwer Academic Publishers: Dordrecht, The Netherlands, 2001; Volume 108, pp. 247–258._22. [CrossRef]
40. Moulinec, H.; Suquet, P. A numerical method for computing the overall response of nonlinear composites with complex microstructure. *Comput. Methods Appl. Mech. Eng.* **1998**, *157*, 69–94. [CrossRef]
41. Lebensohn, R.A. N-site modeling of a 3D viscoplastic polycrystal using fast Fourier transform. *Acta Mater.* **2001**, *49*, 2723–2737. [CrossRef]
42. Eisenlohr, P.; Diehl, M.; Lebensohn, R.A.; Roters, F. A spectral method solution to crystal elasto-viscoplasticity at finite strains. *Int. J. Plast.* **2013**, *46*, 37–53. [CrossRef]
43. Shanthraj, P.; Eisenlohr, P.; Diehl, M.; Roters, F. Numerically robust spectral methods for crystal plasticity simulations of heterogeneous materials. *Int. J. Plast.* **2015**, *66*, 31–45. [CrossRef]
44. Christensen, R.M. Observations on the definition of yield stress. *Acta Mech.* **2008**, *196*, 239–244. [CrossRef]
45. Sachs, G. *Mitteilungen der Deutschen Materialprüfungsanstalten*; Chapter Zur Ableitung einer Fließbedigung; Springer: Berlin/Heidelberg, Germany, 1929; pp. 94–97._12. [CrossRef]
46. Leffers, T.; Van Houtte, P. Calculated and experimental orientation distributions of twin lamellae in rolled brass. *Acta Metall.* **1989**, *37*, 1191–1198. [CrossRef]
47. Ahzi, S.; Asaro, R.J.; Parks, D.M. Application of crystal plasticity theory for mechanically processed BSCCO superconductors. *Mech. Mater.* **1993**, *15*, 201–222. [CrossRef]
48. Molinari, A.; Canova, G.R.; Ahzi, S. A self-consistent approach of the large deformation polycrystal viscoplasticity. *Acta Metall.* **1987**, *35*, 2983–2994. [CrossRef]
49. Lebensohn, R.A.; Tomé, C.N. A self-consistent anisotropic approach for the simulation of plastic deformation and texture development of polycrystals: Application to zirconium alloys. *Acta Metall. Mater.* **1993**, *41*, 2611–2624. [CrossRef]
50. Field, D.P.; Trivedi, P.B.; Wright, S.I.; Kumar, M. Analysis of local orientation gradients in deformed single crystals. *Ultramicroscopy* **2005**, *103*, 33–39. [CrossRef]
51. Kiewel, H.; Bunge, H.J.; Fritsche, L. Effect of the Grain Shape on the Elastic Constants of Polycrystalline Materials. *Textures Microstruct.* **1996**, *28*, 17–33. [CrossRef]
52. Jöchen, K. *Homogenization of the Linear and Non-linear Mechanical behaviour of Polycrystals*; KIT Scientific Publishing: Karlsruhe, Germany, 2013. [CrossRef]
53. Crumbach, M.; Pomana, G.; Wagner, P.; Gottstein, G. A Taylor Type Deformation Texture Model Considering Grain Interaction and Material Properties. Part I—Fundamentals. In *Recrystallisation and Grain Growth, Proceedings of the First Joint Conference*; Gottstein, G.; Molodov, D.A., Eds.; Springer: Berlin, Germany, 2001; pp. 1053–1060.
54. Van Houtte, P. Deformation texture prediction: From the Taylor model to the advanced Lamel model. *Int. J. Plast.* **2005**, *21*, 589–624. [CrossRef]
55. Eisenlohr, P.; Tjahjanto, D.D.; Hochrainer, T.; Roters, F.; Raabe, D. Texture Prediction from a Novel Grain Cluster-Based Homogenization Scheme. *Int. J. Mater. Form.* **2009**, *2*, 523–526. [CrossRef]
56. Tjahjanto, D.D.; Eisenlohr, P.; Roters, F. Relaxed Grain Cluster (RGC) Homogenization Scheme. *Int. J. Mater. Form.* **2009**, *2*, 939–942. [CrossRef]
57. Tasan, C.C.; Diehl, M.; Yan, D.; Bechtold, M.; Roters, F.; Schemmann, L.; Zheng, C.; Peranio, N.; Ponge, D.; Koyama, M.; et al. An overview of dual-phase steels: Advances in microstructure-oriented processing and micromechanically guided design. *Annu. Rev. Mater. Res.* **2015**, *45*, 391–431. [CrossRef]
58. Appel, F.; Wagner, R. Microstructure and deformation of two-phase γ-titanium aluminides. *Mater. Sci. Eng. R Rep.* **1998**, *22*, 187–268. [CrossRef]

59. Zeghadi, A.; N'guyen, F.; Forest, S.; Gourgues, A.F.; Bouaziz, O. Ensemble averaging stress–strain fields in polycrystalline aggregates with a constrained surface microstructure—Part 1: Anisotropic elastic behaviour. *Philos. Mag.* **2007**, *87*, 1401–1424. [CrossRef]
60. Zeghadi, A.; Forest, S.; Gourgues, A.F.; Bouaziz, O. Ensemble averaging stress–strain fields in polycrystalline aggregates with a constrained surface microstructure—Part 2: Crystal plasticity. *Philos. Mag.* **2007**, *87*, 1425–1446. [CrossRef]
61. Zaefferer, S.; Wright, S.I.; Raabe, D. Three-Dimensional Orientation Microscopy in a Focused Ion Beam–Scanning Electron Microscope: A New Dimension of Microstructure Characterization. *Metall. Mater. Trans. A* **2008**, *39*, 374–389. [CrossRef]
62. Wang, L.; Li, M.; Almer, J.; Bieler, T.; Barabash, R. Microstructural characterization of polycrystalline materials by synchrotron X-rays. *Front. Mater. Sci.* **2013**, *7*, 156–169. [CrossRef]
63. Pospiech, J.; Sztwiertnia, K.; Haessner, F. The Misorientation Distribution Function. *Textures Microstruct.* **1986**, *6*, 201–215. [CrossRef]
64. Miodownik, M.; Godfrey, A.W.; Holm, E.A.; Hughes, D.A. On boundary misorientation distribution functions and how to incorporate them into three-dimensional models of microstructural evolution. *Acta Mater.* **1999**, *47*, 2661–2668. [CrossRef]
65. Groeber, M.; Ghosh, S.; Uchic, M.D.; Dimiduk, D.M. A framework for automated analysis and simulation of 3D polycrystalline microstructures. Part 1: Statistical characterization. *Acta Mater.* **2008**, *56*, 1257–1273. [CrossRef]
66. Groeber, M.; Ghosh, S.; Uchic, M.D.; Dimiduk, D.M. A framework for automated analysis and simulation of 3D polycrystalline microstructures. Part 2: Synthetic structure generation. *Acta Mater.* **2008**, *56*, 1274–1287. [CrossRef]
67. Groeber, M.A.; Jackson, M.A. DREAM.3D: A Digital Representation Environment for the Analysis of Microstructure in 3D. *Integr. Mater. Manuf. Innov.* **2014**, *3*, 5. [CrossRef]
68. Diehl, M.; Groeber, M.; Haase, C.; Molodov, D.A.; Roters, F.; Raabe, D. Identifying Structure–Property Relationships Through DREAM.3D Representative Volume Elements and DAMASK Crystal Plasticity Simulations: An Integrated Computational Materials Engineering Approach. *JOM* **2017**, *69*, 848–855. [CrossRef]
69. Ma, A.; Roters, F.; Raabe, D. A dislocation density based consitutive law for BCC materials in crystal plasticity FEM. *Comput. Mater. Sci.* **2007**, *39*, 91–95. [CrossRef]
70. Wong, S.L.; Madivala, M.; Prahl, U.; Roters, F.; Raabe, D. A crystal plasticity model for twinning- and transformation-induced plasticity. *Acta Mater.* **2016**, *118*, 140–151. [CrossRef]
71. Cereceda, D.; Diehl, M.; Roters, F.; Raabe, D.; Perlado, J.M.; Marian, J. Unraveling the temperature dependence of the yield strength in single-crystal tungsten using atomistically-informed crystal plasticity calculations. *Int. J. Plast.* **2016**, *78*, 242–265. [CrossRef]
72. Reuber, C.; Eisenlohr, P.; Roters, F.; Raabe, D. Dislocation density distribution around an indent in single-crystalline nickel: Comparing nonlocal crystal plasticity finite element predictions with experiments. *Acta Mater.* **2014**, *71*, 333–348. [CrossRef]
73. Shanthraj, P.; Sharma, L.; Svendsen, B.; Roters, F.; Raabe, D. A phase field model for damage in elasto-viscoplastic materials. *Comput. Methods Appl. Mech. Eng.* **2016**, *312*, 167–185. [CrossRef]
74. Shanthraj, P.; Svendsen, B.; Sharma, L.; Roters, F.; Raabe, D. Elasto-viscoplastic phase field modelling of anisotropic cleavage fracture. *J. Mech. Phys. Solids* **2017**, *99*, 19–34. [CrossRef]
75. Papanikolaou, S.; Thibault, J.; Woodward, C.; Shanthraj, P.; Roters, F. Brittle to Quasi-Brittle Transition and Crack Initiation Precursors in Disordered Crystals. *arXiv* **2017**, arXiv:1707.04332v1.

© 2019 by the authors. Licensee MDPI, Basel, Switzerland. This article is an open access article distributed under the terms and conditions of the Creative Commons Attribution (CC BY) license (http://creativecommons.org/licenses/by/4.0/).

Article

Analysis of Ductile Fracture Obtained by Charpy Impact Test of a Steel Structure Created by Robot-Assisted GMAW-Based Additive Manufacturing

Ali Waqas, Xiansheng Qin, Jiangtao Xiong, Chen Zheng * and Hongbo Wang

School of Mechanical Engineering, Northwestern Polytechnical University, Xi'an 710072, China; aliwaqas@mail.nwpu.edu.cn (A.W.); xsqin@nwpu.edu.cn (X.Q.); xiongjiangtao@nwpu.edu.cn (J.X.); wanghongbo@nwpu.edu.cn (H.W.)
* Correspondence: chen.zheng@nwpu.edu.cn; Tel.: +86-151-0290-4023

Received: 23 October 2019; Accepted: 7 November 2019; Published: 10 November 2019

Abstract: In this study, gas metal arc welding (GMAW) was used to construct a thin wall structure in a layer-by-layer fashion using an AWS ER70S-6 electrode wire with the help of a robot. The Charpy impact test was performed after extracting samples in directions both parallel and perpendicular to the deposition direction. In this study, multiple factors related to the resulting absorbed energy have been discussed. Despite being a layered structure, homogeneous behavior with acceptable deviation was observed in the microstructure, hardness, and fracture toughness of the structure in both directions. The fracture is extremely ductile with a dimpled fibrous surface and secondary cracks. An estimate for fracture toughness based on Charpy impact absorbed energy is also given.

Keywords: Charpy impact test; GMAW; additive manufacturing; secondary cracks

1. Introduction

Additive manufacturing can be used to create a near-net shape for complex parts using the layer-by-layer deposition method. Powder or wire is melted using different energy sources, including electron beam, laser beam, or electric arc [1–3]. Integrated machinery, such as computer numerical control gantries or robots, can be used to create parts using wire and arc additive manufacturing. The mechanical properties of the manufactured materials generally depend on the welding parameters selected—they have been shown to have better properties than casted materials [4,5]. Researchers have studied different techniques for the process, including conventional gas tungsten arc welding (GTAW), gas metal arc welding (GMAW), and cold metal transfer (CMT) [6,7]. This includes studies on topology, build-up geometry, and material properties of structures made by these methods [8,9]. Comprehensive studies have been conducted on defects in microstructure and methods to improve them by controlling deposition strategies and incorporating ancillary processes for quality enhancement [10–12]. A lot of research is being carried out to control problems related to GMAW-based additive manufacturing, including dimension control at the start and end of the weld bead [13,14]. The height difference at the extreme ends is significant for multi-layer single-pass manufacturing, where the height difference is exaggerated with each layer being deposited, terminating the welding process [15]. Different techniques have been used to control the welding parameters and attain a maximum effective area in the resulting structure [13,16]. The resulting structure has different mechanical and material properties owing to the heat cycles of multiple layer depositions [17].

While studies have been conducted on mechanical properties, including tensile strength and hardness of materials created by additive manufacturing, little work can be found on the impact toughness of these materials. Toughness is an important characteristic that can help study the ability to absorb energy as well as the ductile or brittle behavior of the structure [18]. Toughness may or may not be anisotropic, based on the welding process, microstructure, and grain size [19,20]. Charpy impact testing is one of the most common methods to measure impact toughness. According to a study, the scatter might be lesser at room temperature, as compared to lower temperatures [21].

The results from the Charpy impact test should be studied in more depth, along with the microstructure and fracture analysis of the test specimens, to validate the absorbed impact energy [22]. As the fractography recognizes the mechanism of material failure, the behavior of crack propagation can identify the reasons for higher or lower energy absorbed by the ductile or brittle material. Ductile fractures have a dimpled surface due to tearing of the material and plastic deformation, while brittle fractures are evident from cleavage facets and almost no plastic deformation [23]. Moreover, in the case of ductile fracture, secondary cracks depict the indication of crack deflection with the absorption of more energy, resulting in better toughness [24].

This research focuses on the impact toughness of components made by GMAW and the possible factors responsible for the absorbed energy. The microstructure of the specimens is discussed, along with the fractography of the specimens, after impact testing. The deformation of the broken samples and intrinsic toughening mechanism are discussed in relation to the absorbed energy. An estimate of fracture toughness is also presented.

2. Method and Experiment

A thin wall was constructed by robot-assisted GMAW after controlling the welding parameters at the onset and end of the weld bead. The onset of the weld bead will be referred to as arc-striking, the end will be termed as arc-extinguishing, and the central part will be referred to as the steady stage; a schematic diagram is given in Figure 1. Low carbon steel electrode wire ER70S-6 with a diameter of 1.2 mm has been used to carry out the experiments with the following composition (Table 1).

Table 1. Typical chemical composition for electrode wire ER70S-6 (weight percentage).

Elements	C	Mn	Si	S	P	Ni	Cr	Mo	V	Cu	Fe
wt. %	0.1	1.56	0.88	0.012	0.011	0.01	0.02	<0.01	<0.01	0.24	Bal.

The welding parameters have been controlled on the basis of welding energy profile optimization for uniform height throughout the weld bead. The travel speed is reduced as the weld bead approaches a steady stage to control the bulging shape at the arc-striking region. The decreasing slope at the arc-extinguishing area is controlled by reducing all of the parameters, including current, voltage, and travel speed. In the current study, the samples were extracted from the steady stage with constant welding energy of 660 J/mm and a two-minute delay before deposition of subsequent layers to prevent the process from terminating due to pool flow. This part of the deposition offers equilibrium in terms of height of the deposition and heat dissipation. The details of the same can be found in published literature mentioned in [16]. Deposition parameters for the steady stage part of the layer after equilibrium are provided in Table 2.

Table 2. Deposition parameters for the steady stage part of the layer.

Parameters	Current	Voltage	Welding Energy	Travel Speed
Value (units)	120 (A)	19 (V)	660 (J/mm)	3.5 (mm/s)

Absorbed energy was obtained at room temperature using an automatic impact testing machine JBS-300 (Jinan Kehui Testing Instrument Co., Ltd., Jinan, China) with a maximum capacity of 300 J, as shown in Figure 2. The pre-lift angle was 150°, while the impact velocity was 5.2 m/s. The Brinell hardness test was performed using Huayin 320HBS-3000 (Laizhou Huayin Testing Company Limited, Laizhou, China) to check the macro hardness of the specimens. Optical microscopy was carried out using OLYMPUS GX-71 (Olympus Corporation, Tokyo, Japan), while scanning electron microscopy and fractography was conducted using TESCAN VEGA (Oxford Instruments Technology, Beijing, China). X-ray diffraction (XRD) was performed using X'Pert PRO (PANalytical, Eindhoven, Netherlands) with a copper anode and generator settings of 40 mA and 40 KV.

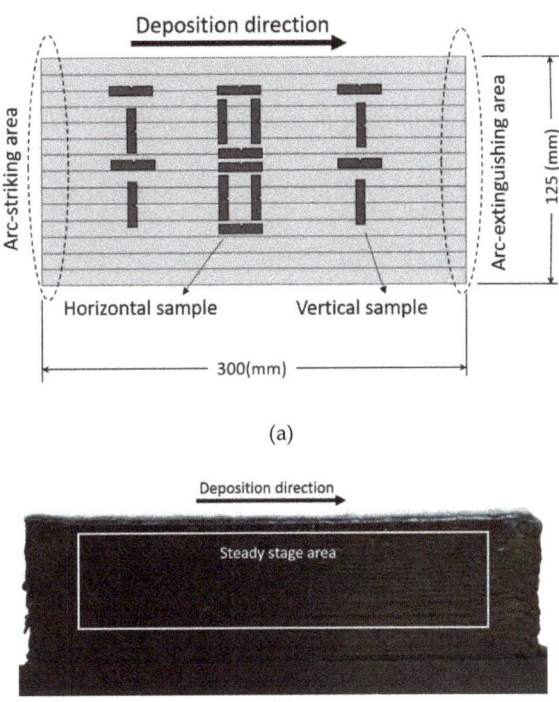

Figure 1. (a) Schematic representation of the robot-assisted welded thin wall, highlighting important areas and the direction of specimen extraction. (b) An as-built wall with depiction of the steady stage from where samples were extracted for this study.

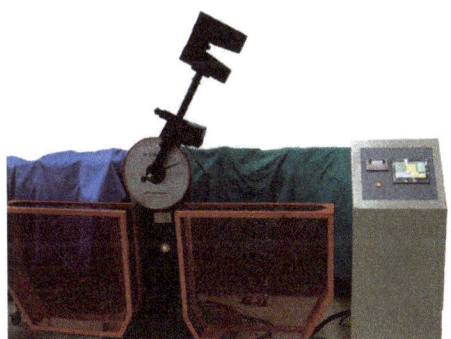

Figure 2. Charpy impact testing machine with a hammer having pre-lift angle of 150°.

The samples were obtained in directions both parallel to the deposition (hereafter referred to as horizontal) and perpendicular to the deposition (hereafter referred to as vertical). Due to the limitation of the available thickness of the thin wall structure, sub-size samples were extracted with dimensions of

$55 \times 10 \times 5$ mm^3 in accordance with the specifications mentioned in standard test methods for notched bar impact testing of metallic materials [25]. Eight samples were extracted in both horizontal and vertical directions. The extracted sample, along with the impact direction used, is shown in Figure 3.

Figure 3. Placement of sample and impact direction.

3. Results and Discussion

Absorbed energy results for both horizontal and vertical samples are given in Figure 4, along with a comparison of absorbed energy of steel with similar carbon content (i.e., 0.11% C) [26]. The values for the absorbed energy have been normalized for the full-size sample. The explanation for the normalizing is provided later in the article. The average value for horizontal specimens (X1 to X8) is approximately 189 J, while it is approximately 202 J for the vertical specimens (Y1 to Y8).

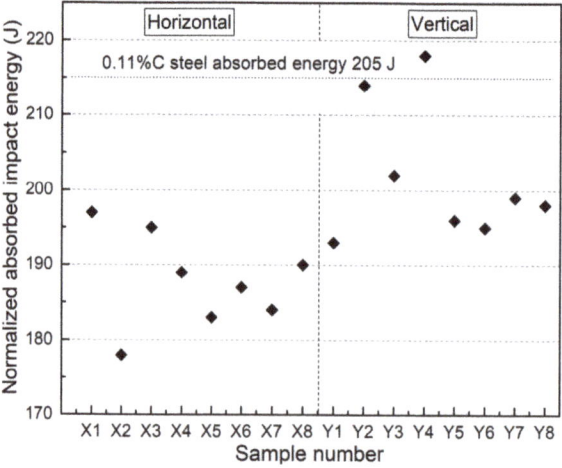

Figure 4. Absorbed impact energy in joules for horizontal and vertical specimens.

Various methods can be adopted to increase the strength of materials, including cold work hardening, precipitate or dispersion hardening, and grain refinement [27]. Strengthening is identified by obstruction in lattice dislocations in all the cases. However, cold working and precipitation hardening increase the brittleness of the material, while grain refinement has a different effect that enhances ductility in terms of percentage elongation [28]. The pre- and post-heat effect of each layer being welded results in the refinement of the grains, causing a higher percentage elongation. The amount of energy absorbed is comparable to the upper shelf absorbed energy of ferritic structure from the reference [26], portraying the material's ability to undergo a large amount of plastic deformation, hence good ductility. Grain structure is depicted in Figure 5 for different directions and magnification levels.

Although the structure of the grains is similar within the layer and between the two successive layers, a difference in size can be observed, as shown in Figure 5a. Higher magnification images are presented in Figure 5b,c for the intralayer and interlayer microstructures, respectively. The same difference is also visible in SEM images in Figure 6a,b. Histogram for a part of each SEM image is given in Figure 6c,d for intralayer and interlayer grain diameter, respectively. As the samples have been taken from the steady stage area where equilibrium has been achieved, the grain structure is mostly ferritic equiaxed, as shown in Figure 5. The average grain size number was found to be 10.5, calculated following ASTM standard E112 [29].

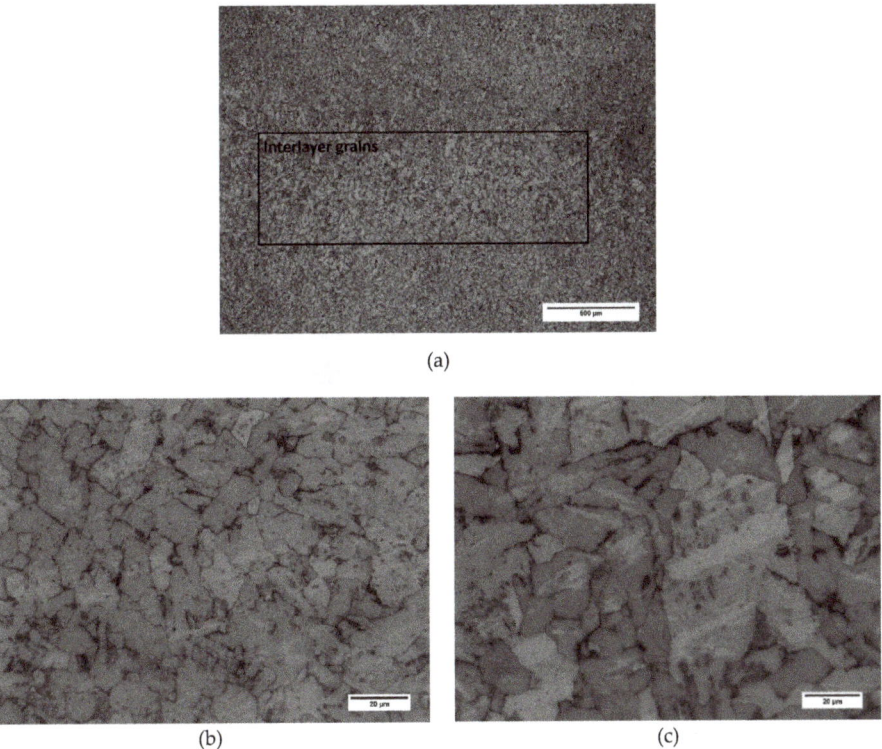

Figure 5. Microstructure attained by an optical microscope. (**a**) Intralayer microstructure with slightly more refined grains. Microstructure with a higher magnification for (**b**) intralayer and (**c**) interlayer grains.

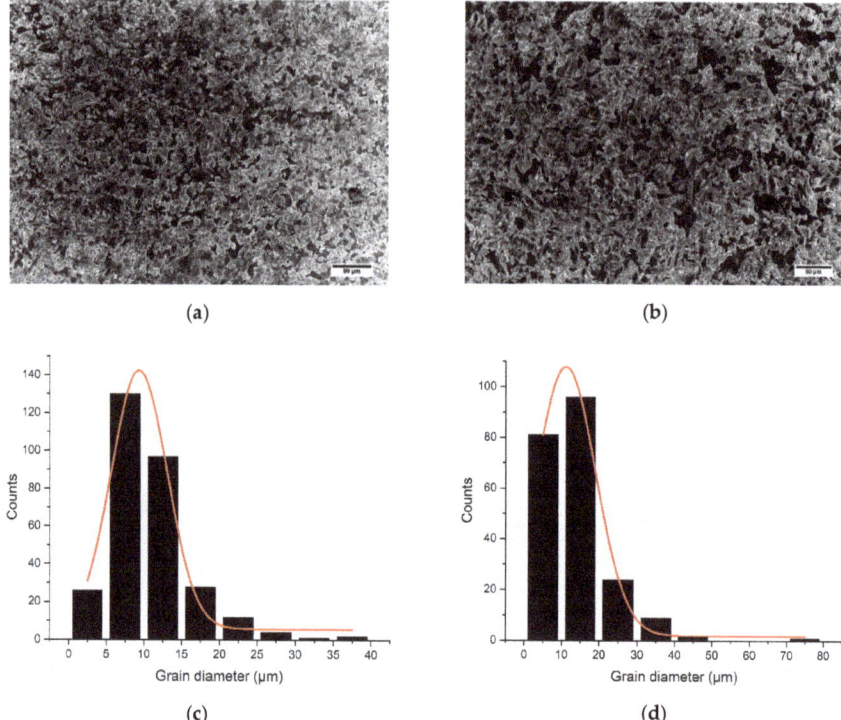

Figure 6. Microstructure attained by an SEM with a slight difference in grain size for the (**a**) intralayer and (**b**) interlayer microstructure. Histogram of grain diameter for the (**c**) intralayer and (**d**) interlayer microstructure.

Although hardness varies in direct proportion to the carbon content, fine grain size results in higher values of hardness [30]. The average hardness value of horizontal samples is approximately 149 BHN with a standard deviation of 1.35, while it is approximately 148.7 BHN with a standard deviation of 0.71 for vertical samples, as shown in Figure 7. Values are comparable to steel with similar carbon content.

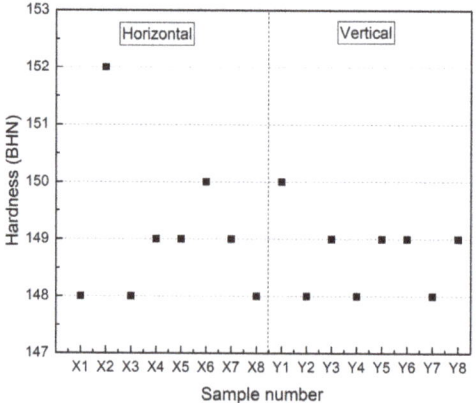

Figure 7. Brinell hardness for horizontal and vertical specimens.

Low carbon steel with a uniform microstructure is considered to have better toughness because of its ferritic structure. High carbon steel with a martensitic structure has more brittle sites, providing lesser resistance to crack propagation [31,32]. However, in this case, a uniform microstructure with a mostly ferritic structure proves to be a hindrance to dislocations in all directions, resulting in a slanted fracture in each plane, as shown in Figure 8. This slanted plane was identical in both the horizontal and vertical specimens, proving that the structure is uniform in both directions with decent penetration of each layer into the subsequent one.

Figure 8. Fractured sample after separation by joining the hinges (three views of the same sample).

The energy absorbed by the material in the plastic region is of importance, especially for structural steel; thus, maximum strength can be estimated by observing deformation ability before the final fracture in the inelastic region. Figure 9 shows the amount of deformation that each sample has undergone before the final fracture. Regardless of deposition direction, each sample has been deformed in a similar fashion with quite a large deformation before failure. This t includes lateral expansion of the specimen, which has been normalized for the sub-size sample (Figure 10). Average values for the horizontal and vertical specimens are approximately 46% and 51%, respectively.

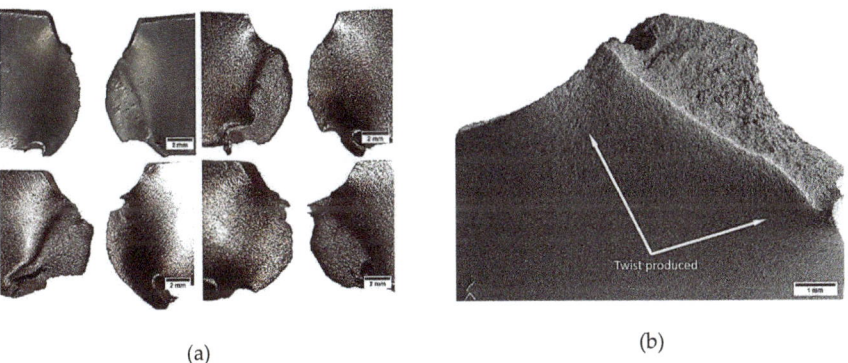

(a)   (b)

Figure 9. (**a**) Multiple samples with identical fracture behavior. (**b**) SEM to show the twist in the deformed sample.

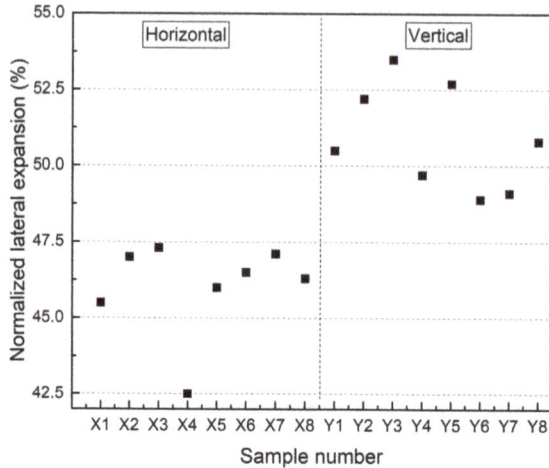

Figure 10. Normalized lateral expansion in percentage for the horizontal and vertical specimens.

The deformation behavior with an almost 45° fracture plane depicts shear stress that exceeds the shear strength of the structure, resulting in plastic yielding. The fibrous appearance and large deformation before fracture, along with the shear plane fracture, point towards the pure ductile fracture. The specimens had to be separated after being closed once at the hinges, according to instructions set by standard test methods for notched bar impact testing of metallic materials ASTM E23-07a [25]. There was no cleavage in the broken specimen; thus, it is considered a pure shear fracture, according to the mentioned standard. The coalescence of voids results in the development of a shear lip, which is responsible for a higher upper shelf energy fracture. Fractography of the specimens displays a dimpled surface with a fibrous fracture, as shown in Figure 11. Generally, brittle fracture in carbon steel is initiated by martensitic sites; however, the microstructure shown in Figure 5 shows that the structure obtained in this case is equiaxed and mostly ferritic. The fractography depicts pure ductile behavior, the reason why the results have been normalized by a factor of two, as presented in Figure 4.

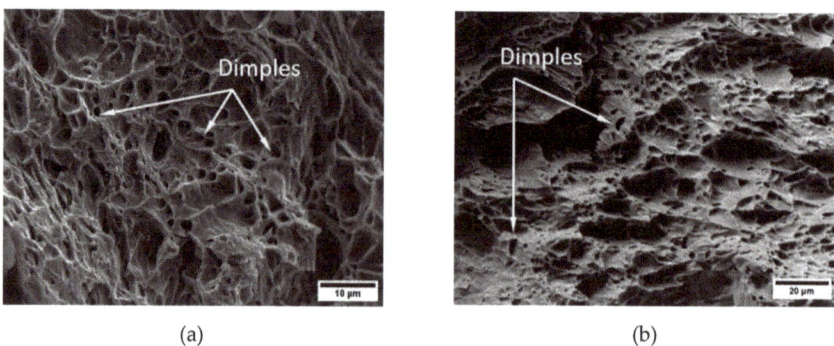

(a)   (b)

Figure 11. Dimpled fracture surface with a fibrous tearing, showing it to be ductile. (a) Horizontal specimen, (b) vertical specimen.

The X-ray diffraction analysis supports the presence of a ferritic structure with mostly pure iron in the central part of the constructed wall, as shown in Figure 12. The trace elements were mostly evaporated or dragged to the extreme ends of the wall. In this kind of fully ferritic structure, microvoid nucleations generate at the grain boundary and deep equiaxed dimples are formed [33].

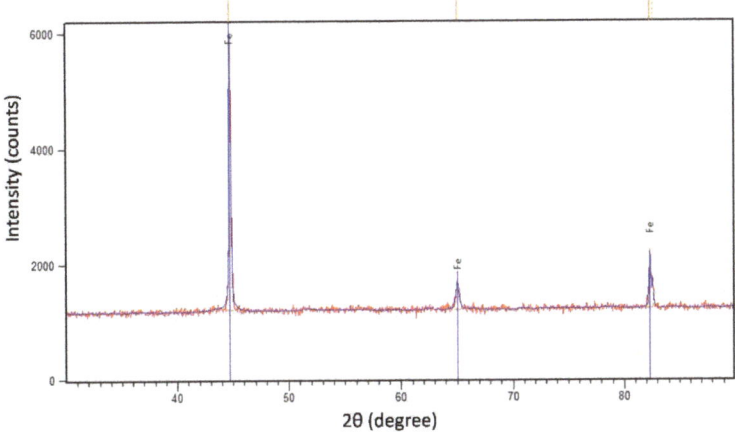

Figure 12. XRD pattern.

Another important factor responsible for the high absorbed energy is the formation of secondary cracks. The creation and motion of dislocations in the crystal lattice are responsible for the plastic deformation. The material dissipates energy during the dislocation movements and crack tip dislocation nucleation leads to intrinsic ductility [34]. The secondary cracks might also have been generated due to the stacked layers, which act as a crack divider, as depicted in Figure 13. This delamination phenomenon can occur even without a substantial difference between the interlayer and intralayer microstructures [35].

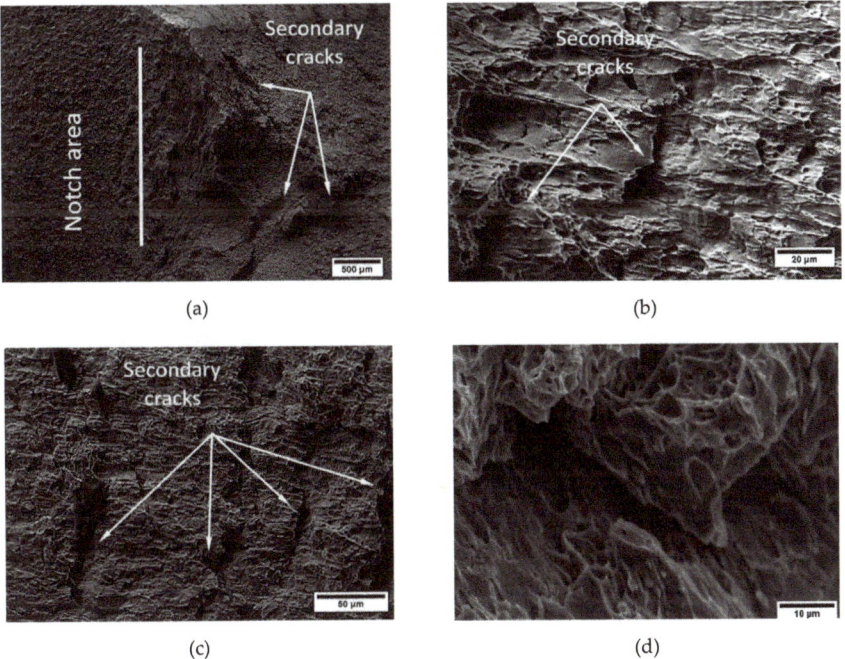

Figure 13. (**a**) Macro-level secondary cracks, (**b**,**c**) micro-level secondary cracks, (**d**) enlarged view of secondary cracks.

As the fracture was purely in the upper shelf region and was ductile in nature, fracture toughness can be estimated using the relation [36,37]:

$$K_{IC} = 0.804\ \sigma_{ys}\ (CVN/\sigma_{ys} - 0.0098)^{0.5} \quad (1)$$

where K_{IC} is the fracture toughness in MPa·m$^{1/2}$, σ_{ys} is the yield strength in MPa, and CVN is the Charpy impact absorbed energy in J. Using the average yield strength (330 MPa) of the structure from [16], the fracture toughness was found to be approximately 199 MPa·m$^{1/2}$ and 206 MPa·m$^{1/2}$ for the horizontal and vertical specimens, respectively.

4. Conclusion

This study presents an analysis to explain the different factors related to the Charpy impact energy absorbed by a structure made by GMAW additive manufacturing. The average absorbed energy in the horizontal and vertical direction was found to be 189 J and 202 J, respectively. The difference in the amount of energy in both directions is not substantial, which is also in conformance with the observed microstructure. The microstructure was found to be mostly equiaxed with a grain size number of about 10.5. The broken samples exhibit a large amount of deformation in all directions, thus absorbing a high amount of energy. Fractography of the broken samples reveals a highly fibrous fracture with dimples, suggesting a pure ductile fracture. The generation of secondary cracks is also indicative of high absorbed energy. As the fracture is in the upper shelf region, the estimated value for fracture toughness was calculated to be 199 MPa·m$^{1/2}$ and 206 MPa·m$^{1/2}$ for the horizontal and vertical specimens, respectively.

Author Contributions: Conceptualization, J.X.; funding acquisition, C.Z.; project administration, X.Q.; resources, H.W.; writing—original draft, A.W.

Funding: This research was financially supported by the National Natural Science Foundation of China, grant number 51805437.

Conflicts of Interest: The authors declare no conflict of interest.

References

1. Frazier, E.W. Metal Additive Manufacturing: A Review. *J. Mater. Eng. Perform.* **2014**, *23*, 1917–1928. [CrossRef]
2. Antonysamy, A. Microstructure, texture and mechanical property evolution during additive manufacturing of Ti6Al4V alloy for aerospace applications. PhD Thesis, University of Manchester, Manchester, UK, 2012.
3. Buckner, M.; Lonnie, J. Automating and accelerating the additive manufacturing design process with multi-objective constrained evolutionary optimization and HPC/Cloud computing. In Proceedings of the 2012 Future of Instrumentation International Workshop (FIIW), Gatlinburg, TN, USA, 8–9 October 2012.
4. Brandl, E.; Baufeld, B.; Leyens, C. Additive manufactured Ti-6Al-4V using welding wire: comparison of laser and arc beam deposition and evaluation with respect to aerospace material specifications. *Physics Procedia* **2010**, *5(Part B)*, 595–606. [CrossRef]
5. Pandremenos, J.; Doukas, C.; Stavropoulos, P.; Chryssolouris, G. Machining with robots: a critical review. In Proceedings of the 7th International Conference on Digital Enterprise Technology, Athens, Greece, 28–30 September 2011.
6. Wang, Y.; Chen, X.; Konovalov, S. Additive manufacturing based on welding arc: a low-cost method. *J. Surf. Invest.* **2017**, *11*, 1317–1328. [CrossRef]
7. Chen, X.; Su, C.; Wang, Y.; Siddiquee, A.N.; Sergey, K.; Jayalakshmi, S.; Singh, R.A. Cold Metal Transfer (CMT) Based Wire and Arc Additive Manufacture (WAAM) System. *J. Surf. Invest.* **2018**, *12*, 1278–1284. [CrossRef]
8. Müller, J.; Grabowski, M.; Müller, C.; Hensel, J.; Unglaub, J.; Thiele, K.; Kloft, H.; Dilger, K. Design and Parameter Identification of Wire and Arc Additively Manufactured (WAAM) Steel Bars for Use in Construction. *Metals* **2019**, *9*, 725. [CrossRef]

9. Rodrigues, T.A.; Duarte, V.; Avila, J.A.; Santos, T.G.; Miranda, R.; Oliveira, J. Wire and arc additive manufacturing of HSLA steel: Effect of thermal cycles on microstructure and mechanical properties. *Addit. Manuf.* **2019**, *27*, 440–450. [CrossRef]
10. Oliveira, J.; Santos, T.; Miranda, R. Revisiting fundamental welding concepts to improve additive manufacturing: From theory to practice. *Prog. Mater. Sci.* **2019**, *107*, 100590. [CrossRef]
11. Rodrigues, T.A.; Duarte, V.; Miranda, R.; Santos, T.G.; Oliveira, J. Current Status and Perspectives on Wire and Arc Additive Manufacturing (WAAM). *Materials* **2019**, *12*, 1121. [CrossRef]
12. Cunningham, C.; Flynn, J.; Shokrani, A.; Dhokia, V.; Newman, S. Invited review article: Strategies and processes for high quality wire arc additive manufacturing. *Addit. Manuf.* **2018**, *22*, 672–686. [CrossRef]
13. Xiong, J.; Yin, Z.Q.; Zhang, W.H. Forming appearance control of arc striking and extinguishing area in multi-layer single-pass GMAW-based additive manufacturing. *Int. J. Adv. Manuf. Tech.* **2016**, *87*, 579–586. [CrossRef]
14. Hu, Z.; Qin, X.; Shao, T.; Liu, H. Understanding and overcoming of abnormity at start and end of the weld bead in additive manufacturing with GMAW. *Int. J. Adv. Manuf. Tech.* **2018**, *95*, 2357–2368. [CrossRef]
15. Zhang, Y.; Chen, Y.; Li, P.; Male, A.T. Weld deposition-based rapid prototyping: a preliminary study. *J. Mater. Process. Technol.* **2003**, *135*, 347–357. [CrossRef]
16. Waqas, A.; Qin, X.; Xiong, J.; Wang, H.; Zheng, C. Optimization of Process Parameters to Improve the Effective Area of Deposition in GMAW-Based Additive Manufacturing and its Mechanical and Microstructural Analysis. *Metals* **2019**, *9*, 775. [CrossRef]
17. Zhao, H.H.; Zhang, G.J.; Yin, Z.Q.; Wu, L. A 3D dynamic analysis of thermal behavior during single-pass multi-layer weld-based rapid prototyping. *J. Mater. Process. Technol.* **2011**, *211*, 488–495. [CrossRef]
18. Cao, W.; Zhang, M.; Huang, C.; Xiao, S.; Dong, H.; Weng, Y. Ultrahigh Charpy impact toughness (~450 J) achieved in high strength ferrite/martensite laminated steels. *Sci. Rep.* **2017**, *7*, 41459. [CrossRef]
19. Rojas, P.; Martinez, C.; Vera, R.; Puentes, M. Toughness of SAE 1020 Steel with and without Galvanization Exposed to Different Corrosive Environments in Chile. *Int. J. Electrochem. Sci.* **2014**, *9*, 2848–2858.
20. Salimi, A.; Zadeh, H.M.; Toroghinejad, M.R.; Asefi, D.; Ansaripour, A. Influence of Sample Direction on the Impact Toughness of the Api-X42 Microalloyed Steel with a Banded Structure. *Mater. Technol..* **2013**, *47*, 385–389.
21. Lucon, E.; McCowan, C.N.; Santoyo, R.L. Certification of NIST Room Temperature Low-Energy and High-Energy Charpy Verification Specimens. *J. Res. Nat. Inst. Stand. Technol.* **2015**, *120*, 316–328. [CrossRef]
22. Dean, S.W.; Manahan, M.P.; Mccowan, C.N. Percent Shear Area Determination in Charpy Impact Testing. *J. ASTM Int.* **2008**, *5*, 101662–101676. [CrossRef]
23. Parrington, R.J. Fractographic features in metals and plastics. *Adv. Mater. Processes* **2003**, *161*, 37–40.
24. Wei, R.P. *Fracture mechanics: Integration of mechanics, materials science and chemistry*; Cambridge University Press: Cambridge, UK, 2010; pp. 24–54.
25. *ASTM E23: Standard Test Methods for Notched Bar Impact Testing of Metallic Materials*; ASTM International: West Conshohocken, PA, USA, 2007.
26. Davis, J. *Metals Handbook, Desk, Edition*, 2nd ed.; ASM International, The Materials Information Society: West Conshohocken, PA, USA, 1998; pp. 560–564.
27. Arzt, E. Size effects in materials due to microstructural and dimensional constraints: a comparative review. *Acta Mater.* **1998**, *46*, 5611–5626. [CrossRef]
28. Calcagnotto, M.; Adachi, Y.; Ponge, D.; Raabe, D. Deformation and fracture mechanisms in fine-and ultrafine-grained ferrite/martensite dual-phase steels and the effect of aging. *Acta. Mater.* **2011**, *59*, 658–670. [CrossRef]
29. *ASTM E112: Standard Test Methods for Determining Average Grain Size*; ASTM International: West Conshohocken, PA, USA, 1996; pp. 4–20.
30. Gharibshahiyan, E.; Raouf, A.H.; Parvin, N.; Rahimian, M. The effect of microstructure on hardness and toughness of low carbon welded steel using inert gas welding. *Mater. Des.* **2011**, *32*, 2042–2048. [CrossRef]
31. Zhao, Y.; Tong, X.; Wei, X.; Xu, S.; Lan, S.; Wang, X.-L.; Zhang, Z. Effects of microstructure on crack resistance and low-temperature toughness of ultra-low carbon high strength steel. *Int. J. Plast.* **2019**, *116*, 203–215. [CrossRef]

32. Thompson, S. Interrelationships between yield strength, low-temperature impact toughness, and microstructure in low-carbon, copper-precipitation-strengthened, high-strength low-alloy plate steels. *Mater. Sci. Eng. A* **2018**, *711*, 424–433. [CrossRef]
33. Kocatepe, K.; Cerah, M.; Erdogan, M. The tensile fracture behaviour of intercritically annealed and quenched+ tempered ferritic ductile iron with dual matrix structure. *Mater. Des.* **2007**, *28*, 172–181. [CrossRef]
34. Kysar, J.W. Energy dissipation mechanisms in ductile fracture. *J. Mech. Phys. Solids* **2003**, *51*, 795–824. [CrossRef]
35. Das, A.; Viehrig, H.-W.; Altstadt, E.; Bergner, F.; Hoffmann, J. Why Do Secondary Cracks Preferentially Form in Hot-Rolled ODS Steels in Comparison with Hot-Extruded ODS Steels? *Crystals* **2018**, *8*, 306. [CrossRef]
36. Chao, Y.J.; Ward Jr, J.; Sands, R.G. Charpy impact energy, fracture toughness and ductile-brittle transition temperature of dual-phase 590 Steel. *Mater. Des.* **2007**, *28*, 551–557. [CrossRef]
37. Barsom, J.M.; Rolfe, S.T. *Fracture and Fatigue Control in Structures: Applications of Fracture Mechanics*; ASTM International: West Conshohocken, PA, USA, 1977; pp. 121–132.

© 2019 by the authors. Licensee MDPI, Basel, Switzerland. This article is an open access article distributed under the terms and conditions of the Creative Commons Attribution (CC BY) license (http://creativecommons.org/licenses/by/4.0/).

Article

Strain Rate Contribution due to Dynamic Recovery of Ultrafine-Grained Cu–Zr as Evidenced by Load Reductions during Quasi-Stationary Deformation at 0.5 T_m

Wolfgang Blum [1,*], Jiři Dvořák [2], Petr Král [2], Philip Eisenlohr [3] and Vaclav Sklenička [2]

[1] Department of Materials Science, Institute I, University of Erlangen-Nuremberg, Martensstr. 5, D-91058 Erlangen, Germany
[2] Institute of Physics of Materials, Czech Academy of Sciences, Žižkova 22, CZ-616 62 Brno, Czech Republic; dvorak@ipm.cz (J.D.); pkral@ipm.cz (P.K.); sklen@ipm.cz (V.S.)
[3] Chemical Engineering and Materials Science, Michigan State University, East Lansing, MI 48824, USA; eisenlohr@egr.msu.edu
[*] Correspondence: wolfgang.blum@fau.de

Received: 24 September 2019; Accepted: 21 October 2019; Published: 26 October 2019

Abstract: During quasi-stationary tensile deformation of ultrafine-grained Cu-0.2 mass%Zr at 673 K and a deformation rate of about $10^{-4}\,\text{s}^{-1}$ load changes were performed. Reductions of relative load by more than about 25% initiate anelastic back flow. Subsequently, the creep rate turns positive again and goes through a relative maximum. This is interpreted by a strain rate component $\dot{\varepsilon}^-$ associated with dynamic recovery of dislocations. Back extrapolation indicates that $\dot{\varepsilon}^-$ contributes the same fraction of $(20 \pm 10)\%$ to the quasi-stationary strain rate that has been reported for coarse-grained materials with high fraction of low-angle boundaries; this suggests that dynamic recovery of dislocations is generally mediated by boundaries. The influence of anelastic back flow on $\dot{\varepsilon}^-$ is discussed. Comparison of $\dot{\varepsilon}^-$ to the quasi-stationary rate points to enhancement of dynamic recovery by internal stresses. Subtraction of $\dot{\varepsilon}^-$ from the total rate yields the rate component $\dot{\varepsilon}^+$ related with generation and storage of dislocations; its activation volume is in the order expected from the classical theory of thermal glide.

Keywords: Cu–Zr; ECAP; ultrafine-grained material; deformation; dynamic recovery; transient; load change tests

1. Introduction

In materials science one is used to think in terms of strain hardening and recovery: The dislocation density increases with plastic strain so that the material hardens; recovery decreases this density with time so that the material softens. However, this view is too simple as the recovery processes get biased under stress so that dynamic recovery, i.e., recovery under stress, also causes strain. Recovery by cross slip is an early and well known example. When the rate of strain due to recovery decreases, the material may seem to harden even though it recovers. The present work deals with this surprising effect in an ultrafine-grained material with high content of high-angle boundaries (HABs).

In a companion paper the quasi-stationary (qs) deformation strength of ultrafine-grained (ufg) Cu–Zr has been described. In qs deformation storage and recovery of dislocations approximately balance each other so that the dislocation density ρ remains approximately constant, i.e.,

$$\dot{\rho}^+ \approx \dot{\rho}^-. \tag{1}$$

Storage occurs after expansion of dislocation loops on slip planes (Figure 1: 'dislocations in').
Dynamic recovery is coagulation of dislocation loops after dipole capture (Figure 1: 'dislocations out').
The recovery processes may be spatially concentrated at crystallite boundaries or may be more equally
distributed as in solid solutions of class I-type with solute drag on dislocations [1,2]. Recovery generally
requires dislocation motion outside the primary slip plane by climb or cross slip [3].

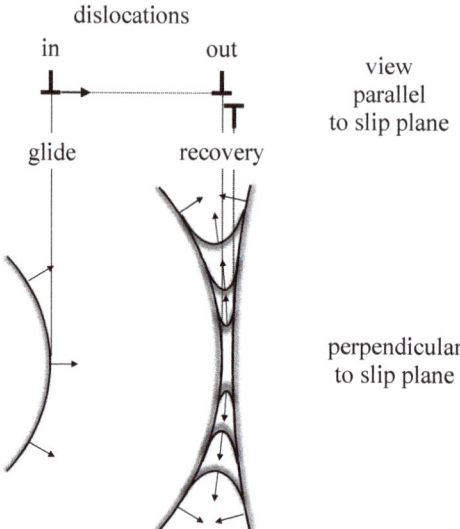

Figure 1. Scheme of dislocation glide with generation and storage of dislocations ('dislocations in') and
dynamic recovery of dislocations ('dislocations out') viewed perpendicular and parallel to glide plane.

In the view in a direction parallel to the glide plane, where dislocations appear as points
(Figure 1), recovery seems to make a negligible contribution to strain during annihilation of dislocation
dipoles. Therefore, dynamic recovery is usually not considered as a process generating strain. Rather,
the models regard strain as a result of thermally activated expansion of slipped areas bounded
by dislocation lines with positive curvature that have to overcome a significant athermal stress
component (forest dislocations, long-range back stresses from boundaries). In this picture the existing
dislocations act as *obstacles* to dislocation glide. However, the view in a direction perpendicular to the
slip plane shows that strain may well be generated during the process of coagulation of dislocation
loops in recovery as negatively curved dislocation segments straighten [4,5]. Here, the interaction of
dislocations *supports* the expansion of slipped areas by glide rather than opposing it. The difference
in driving forces means that the kinetics of generation of dislocation length by glide of positively
curved dislocations moving through the existing dislocation structure differs from that of decrease
of dislocation length by motion of negatively curved dislocations. Therefore, it makes sense to treat
the rate $\dot{\epsilon}_{pl}$ of plastic deformation as sum of storage strain occurring at a rate $\dot{\epsilon}^+$ and recovery–strain
occurring at a rate $\dot{\rho}^-$ [4,5]:

$$\dot{\epsilon}_{pl} = \dot{\epsilon}^+ + \dot{\epsilon}^-. \tag{2}$$

In the literature, there are a couple of examples of processes of type $\dot{\epsilon}^-$, where recovery is coupled
with glide or glide is associated with recovery (class I alloys with viscously moving dislocations [1,2,6],
knitting-out of dislocations from LABs [7–9], accommodation processes at HABs [10], strain coupled
with migration of LABs (e.g., [11]) and HABs (e.g., [12]). Compared to $\dot{\epsilon}^+$ the recovery–strain rate $\dot{\epsilon}^-$
has received little attention (see e.g., [3,13]). In monotonic qs deformation, the two terms $\dot{\epsilon}^+$ and $\dot{\epsilon}^-$ are
coupled via condition Equation 1. To investigate recovery of dislocation lines separately from storage
of dislocations, one must decouple the two processes. This can be done by perturbing monotonic

flow by a sudden change of the force F at which the specimen deforms. Such a perturbation abruptly changes the forces exerted per length of dislocations and triggers reversible time-dependent dislocation motions (e.g., bowing/unbowing). The strains caused by those motions are called anelastic. So the total inelastic strain rate is

$$\dot{\epsilon}_{\text{inel}} = \dot{\epsilon}_{\text{pl}} + \dot{\epsilon}_{\text{anel}}. \tag{3}$$

Figure 2 schematically shows the response to a change from F_0 to $F_r \equiv R F_0$ at a time t_0 and an inelastic strain $\epsilon_{r,0}$. Consider relatively small changes of the relative load R (cases a and b in Figure 2). These cause relatively small changes in inelastic deformation rates from the value $\dot{\epsilon}_{r,0}$ before the R-change to a new value $\dot{\epsilon}_{r,1}$. Anelastic strains are negligible. Just after the R-change, the dislocation structure and the rest of the microstructure are virtually the same as before ('constant structure'), but the glide velocity of dislocations has changed due to the change of the stresses acting on the dislocations. The ratio $\dot{\epsilon}_{r,1}/\dot{\epsilon}_{r,0}$ is widely used to get a measure of the activation volume V^+ of thermally activated dislocation glide as described in more detail in Appendix B. A particularly large body of 'constant structure' data of $\dot{\epsilon}_{r,1}/\dot{\epsilon}_{r,0}$ has been collected for various metals and alloys by Milička in stress change tests during creep at elevated temperatures [14–16].

Now we consider relatively large F-changes (cases c and d in Figure 2). Anelastic strains are no longer negligible and diminish $\dot{\epsilon}_{\text{inel}}$ compared to $\dot{\epsilon}_{\text{pl}}$ (Equation (3)). At sufficiently low R, the forces acting on the dislocations initially get negative so that $\dot{\epsilon}_{\text{inel}}$ becomes negative directly after the R-reduction [17]. This is a consequence of internal stresses of short- and long-range nature acting on the dislocations [17] and opposing thermally activated glide of type $\dot{\epsilon}^+$. As the back flow relaxes the internal back stresses created before the R-reduction, the absolute magnitude of the rate $\dot{\epsilon}_{\text{anel}}$ declines, $\dot{\epsilon}_{\text{pl}}$ becomes dominant again, and forward deformation is reestablished at a rate $\dot{\epsilon}_{\text{pl}} = \dot{\epsilon}_{r,2}$. The preceding anelastic back flow is expected to cause only subtle changes of the dislocation arrangement and the rest of the microstructure; therefore, the rate $\dot{\epsilon}_{r,2}$, measured short after the period of back flow, has also been adressed as 'constant structure' rate. However, it is clear that this is not fully correct (see Equation (4)).

In the further course of the transient after large R-reductions, $\dot{\epsilon}_{\text{anel}}$ becomes negligible so that $\dot{\epsilon}_{\text{inel}} \approx \dot{\epsilon}_{\text{pl}}$. A remarkable result is that $\dot{\epsilon}_{\text{inel}}$ generally *decreases* for long times as schematically indicated by the dashed curve. This behavior is not well known in the community, although it is regularly found whenever investigated, independent of materials and pretreatment. It is distinct from the so-called inverse transient behavior where the decrease of $\dot{\epsilon}_{\text{inel}}$ with strain after R-reduction occurs in the whole interval $0 < R < 1$, and not only at small R. One reason for the lack of knowledge about decreasing $\dot{\epsilon}_{\text{inel}}$ after large R-reductions is, that long-term tests are required for such observations, covering test times distinctly beyond the extended period of back flow. Such tests have been done by Blum and coworkers on a number of materials including e.g., Al–5Mg (class I alloy) [18,19], Al–Zn (class II alloy) [20], and pure LiF [21] and by Van Swygenhoven and coworkers on nanocrystalline Ni and Ni-Fe [10,22,23]. In these tests direct evidence for ongoing net recovery of dislocations was obtained. A natural explanation of the decrease of $\dot{\epsilon}_{\text{inel}}$ after perturbation of plastic flow by a large R-reduction is that the recovery rate component $\dot{\epsilon}^-$ decreases, because the driving force for recovery declines during the decrease of ρ and other crystal defects to the lower level in the new qs state at the lower stress.

The same process of net recovery must also be expected when a deformed specimen is simply unloaded to $R = 0$ and subsequently annealed at elevated temperature higher than the deformation temperature. This type of experiment has been done by Hasegawa, Yakou and Kocks on pure Al [24,25] that was deformed at ambient temperature and then quickly heated to elevated temperature. The result was qualitatively the same as the result of unloading at fixed temperature described before: net *back* flow due to anelastic strains was followed by net *forward* flow at declining rate. This forward flow at zero stress after predeformation was interpreted by the authors as consequence of recovery; the recovery was suggested to result from reaction of neighboring polarized dislocation walls.

So far, comprehensive studies of the transient response to stress reductions are missing in the case of ECAP-processed ufg materials. Two tests performed on ufg Cu [26] showed a decrease of creep rate after relative stress reductions to $R = 0.77$ and 0.70 that could be explained in terms of decreasing recovery–strain rate. The present study of transient deformation after qs deformation of ufg Cu–Zr has these main objectives:

- demonstrate that the transient response to load changes can be studied in standard tensile creep machines with load control,
- advertise a new type of plot [27] (Figure 3e) displaying the full strain-time evolutions of all tests of a series with different degrees of unloading at reasonable resolution,
- show that the transient behavior of an ufg material is qualitatively the same as that of cg materials, including an initial period of strain mainly due to recovery,
- discuss the mechanism of dynamic recovery in qs and transient deformation with special regard to the influence of internal stresses.

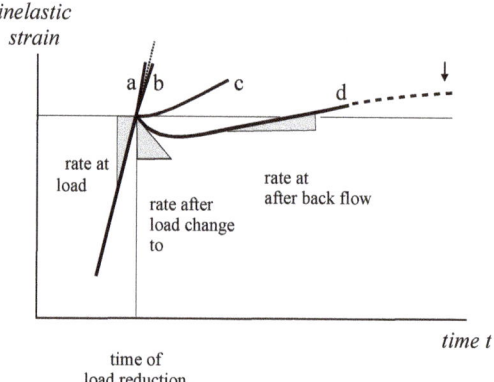

Figure 2. Response of inelastic strain to fast changes of creep load from F_0 to $F_r = R\,F_0$ during deformation at time t_0 and strain $\epsilon_{r,0}$ for (a) small R-increase, (b) small R-decrease, (c) medium R-decrease causing $\dot{\epsilon}_{r,1} = 0$, (d) large R-decrease causing net back flow.

2. Experimental Details

As described in more detail in the companion paper [28], our particle-stabilized material, called pCu–Zr, was produced by severe predeformation at ambient temperature in p passes of equal channel angular pressing (ECAP) on route B_C. Its material parameters are approximated by those of pure Cu provided in the data compilation of Frost and Ashby [29]: Burgers vector $b = 2.56 \times 10^{-10}$ m, elastic shear modulus $G = 3.58 \times 10^4$ MPa, melting point $T_m = 1356$ K. The test temperature was $T = 673$ K $= 0.5\,T_m$.

Deformation was started by applying tensile loads F to flat specimens with initial values of gauge length $l_0 = 10$ mm and cross section S_0 of usually ≈ 12 mm^2. The standard creep machines used in this work were designed for long-term measurements of creep strain accumulation at *constant* load, not for precisely following small strain changes after load changes. The reproducibility of measurements of back flow was worse than in Milička's tests [14–16], but better than originally expected, although some artifacts from unmotivated jumps in the extensometer system or errors in σ_{eng} occasionally seem to have occurred (see e.g., the black curve in Figure 3b after unloading). In the periods of deformation (creep) at constant load the inelastic strain rate is practically identical to the measured total strain rate $\dot{\epsilon}_{tot}$ as the elastic strain rate $\dot{\epsilon}_{el}$ is negligible. In the periods of fast changes of load F this is no longer so. Appendix A explains the procedure taken to get the inelastic strain ϵ_{inel} at acceptable accuracy. The inelastic strain rate follows from ϵ_{inel} as $\dot{\epsilon}_{inel} = \Delta\epsilon_{inel}/\Delta t$ where $\Delta\epsilon_{inel}$ must be chosen larger than the experimental noise. This was achieved by data smoothing with the open software SmooMuDS [30].

3. Results

3.1. Transients as Function of Time

A change of load from a start value F_0 corresponding to an engineering stress $\sigma_{eng} = F_0/S_0$ to a new value $F = R\,F_0$ at time t_0 and inelastic strain ϵ_0 initiates a transient response. To display all transients of largely different durations in the same plot, a logarithmic time scale is used in Figure 3; the constants 10 s in the time-scale and 0.01 in the ϵ_{inel}-scale serve to bring the start of transient into the field of view. Figure 3a–c shows three tests with relative load reductions to by 60% to $R = 0.4$. The reductions deliberately were performed in steps to explore the behavior at intermediate stresses (Figure 3a). The strain evolution varies with step height and step length. In some cases net forward deformation continued during the first unloading steps (Figure 3b). However, the strains accumulated there were small and no significant effect on the values of $\dot{\epsilon}_{inel} > 0$ after the reductions was observed. This is different in the periods of back flow ($\dot{\epsilon}_{inel} < 0$). Such a difference must be expected because back flow relaxes the internal stresses driving it. However, our work does not focus on back the flow triggered by the perturbation by R-reductions, but on the subsequent forward flow (see Figure 3b). Figure 3c displays the *forward* strain rates $\dot{\epsilon}_{inel} > 0$ after R-reduction that reappear after about 20 to 30 ks when back flow has faded, $\dot{\epsilon}_{anel}$ has become negligible and $\dot{\epsilon}_{inel} \approx \dot{\epsilon}_{pl}$. In the beginning, the uncertainty in $\dot{\epsilon}_{inel}$ is large, because relatively small strain intervals $\Delta\epsilon_{inel}$ were used in determination of $\dot{\epsilon}_{inel}$ (compare Equation (2)).

Two of the $\dot{\epsilon}_{inel}$-curves in Figure 3c still appear somewhat noisy. Yet further smoothing of data was avoided because the $\dot{\epsilon}_{inel}$-variations seem to have a real origin in slow T-fluctuations caused by the control system. The two gray curves for 8Cu–Zr in subfigure b show the measured $\dot{\epsilon}_{inel}$-extremes. They differ by a factor of 3 to 4 in $\dot{\epsilon}_{inel}$. We ascribe that to the aforementioned inhomogeneity of the grain structure of 8Cu–Zr. The upper gray curve for 8Cu–Zr is quite similar to the black curve for 12Cu–Zr. We conclude from this result that, apart from the scatter of the initial microstructure produced by the thermomechanical history, there is no significant difference between the ufg materials 8Cu–Zr and 12Cu–Zr.

Figure 3d-f gives the overview of all R-reduction tests performed in this work. Again, we focus on the forward flow observed *after* the anelastic back flow. The curves in Figure 3f derived from Figure 3e are arranged in a fairly consistent sequence corresponding to the loads shown in Figure 3d. This underscores the quality of the length measurements in our creep machines although these were not built for load change tests. For $R \leq 0.3$ a transient decrease of the (forward) strain rate $\dot{\epsilon}_{inel} > 0$ is evident.

Figure 4 shows the times t_{back} (circles) for anelastic back flow taken from the length-time recordings. Due to differences in unloading histories and uncertainties in length measurement the scatter is large. The dashed line corresponds to the dashed curve from Figure 3f approximating the boundary of back flow. For $R > 0.75$ the time interval of back flow is immeasurably small. So back flow becomes negligible here and deformation goes on at positive rate directly after the load reduction.

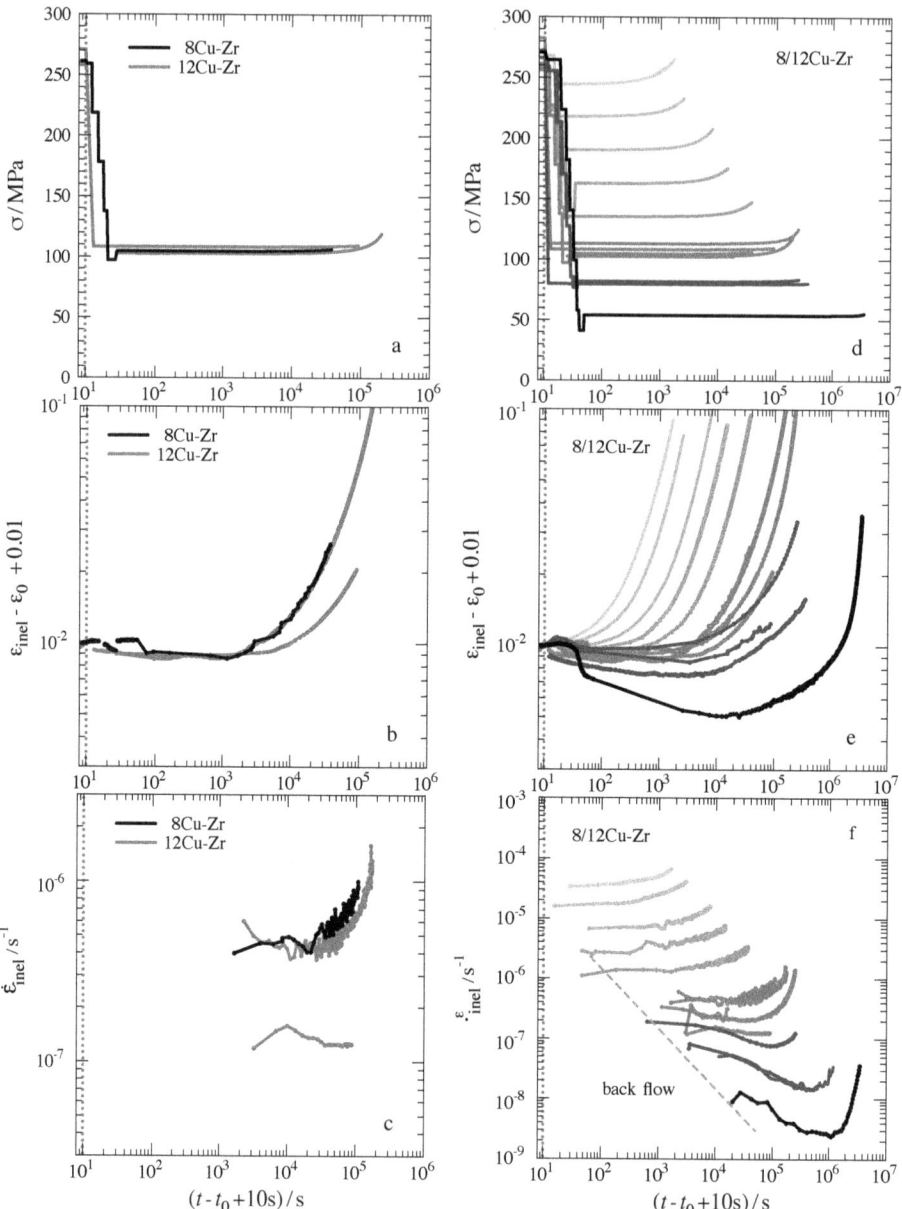

Figure 3. (a) Stress σ, (b) strain ε_{inel}, and (c) strain rate $\dot\varepsilon_{inel}$ as function of time t in tests for 8Cu–Zr and 12Cu–Zr with stepwise load reduction to (**a**–**c**) $R = 0.4$ and (**d**–**f**) all R; dashed line in (**f**) approximates boundary of back flow.

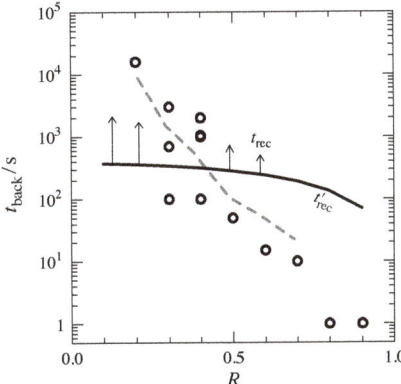

Figure 4. Times t_{back} for anelastic back flow (circles, from Figure 3e) compared to lower bound t'_{rec} of times t_{rec} for dynamic recovery toward the new qs state as function of R for $\sigma_{r,0} = 275$ MPa; dashed line corresponds to dashed line in Figure 3f.

3.2. Transients as Function of Strain

Dislocation generation needs strain. Therefore, the strain ϵ_{inel} is much more closely related to the microstructural evolution than the testing time t. So the evolution of deformation strength $(\sigma, \dot{\epsilon}_{inel})$ is commonly displayed on a strain scale. Figure 5 exhibits the transients of Figure 3f as function of ϵ_{inel}. As σ increases at constant load F, $\dot{\epsilon}_{inel}$ increases even if the microstructure is constant. This effect was eliminated by correcting $\dot{\epsilon}_{inel}$ (see caption). The corrected curves in Figure 5 should be horizontal in the qs state if the grain and phase structure remains constant. This is indeed found for large R near 1. For smaller R the curves exhibit a positive slope in the whole strain interval. This means that slow microstructural changes are going on throughout the test. Comparison of the dotted and the solid curves at $R = 0.4$ and 0.3 shows that these changes are the same in tests with and without R-reduction. At the lowest R of 0.2 (80% unloading) deformation is slowest and the structural changes including dislocations are largest. Consequently, softening is most pronounced here. The curve for $R = 0.2$ was followed for 42 days before it was interrupted without any indications of fracture; note that the $\dot{\epsilon}(\epsilon)$ curve is concave, not convex as in fracture. In [28] the softening has been shown to be a consequence of microstructural coarsening, in particular grain coarsening. This means that only the short-term portions of the curves after R-reduction show the transient response to perturbance of the dynamic equilibrium of storage and recovery of dislocations in the qs state at t_0.

Note that the character of this short-term portion of the transients changes significantly with R. For small R-reductions to $R \geq 0.5$ there is a relative *increase* of $\dot{\epsilon}_{inel}$ compared to the qs curve at reduced R. This is known as normal transient behavior: the material softens due to coarsening of the cellular dislocation structure towards the new dynamic equilibrium state. However, for large R-reductions to $R < 0.5$ and $\dot{\epsilon}_{inel} \leq 10^{-7}$ s^{-1} there is an initial *decrease* of $\dot{\epsilon}_{inel}$.

Figure 6 displays the constant structure rates $\dot{\epsilon}_{r,1}$ and $\dot{\epsilon}_{r,2}$ that were measured at the beginning of the transients and after anelastic back flow, respectively (see Figure 2). Figure 6a shows that $\dot{\epsilon}_{r,1}$ falls to zero near $R = 0.76$ and becomes negative (back flow) for lower R. Following Milicka [14], the data were approximated by a sinh-expression

$$\dot{\epsilon}_{r,1} = k_1 \sinh(V(\sigma - \sigma_i)/(M k_B T)) \qquad k_1 = 0.0885, \sigma_i = 0.76 \sigma_{r,0}, \qquad (4)$$

giving the solid grey line with change from positive to negative (back flow) rates $\dot{\epsilon}_{r,1}$. Figure 6b shows the positive rates $\dot{\epsilon}_{r,2}$ after back flow.

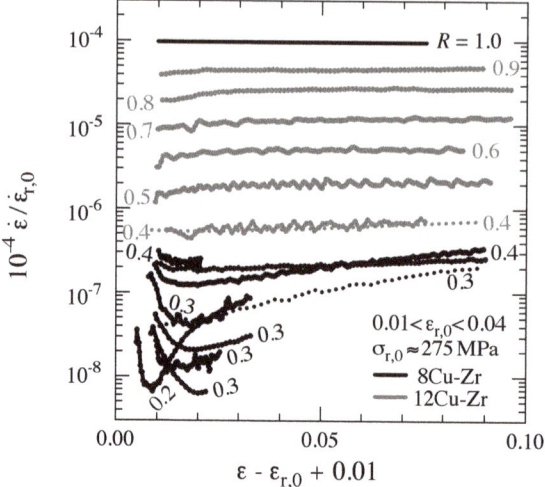

Figure 5. Normalized strain rate as a function of normalized strain after load reduction from $\sigma_{eng} = 250$ MPa to a relative load R for 12Cu–Zr (grey lines) and 8Cu–Zr (black lines); the increase of qs $\dot{\varepsilon} \propto \sigma^{n_{qs}}$ was eliminated with $n_{qs} = 6$ from [28].

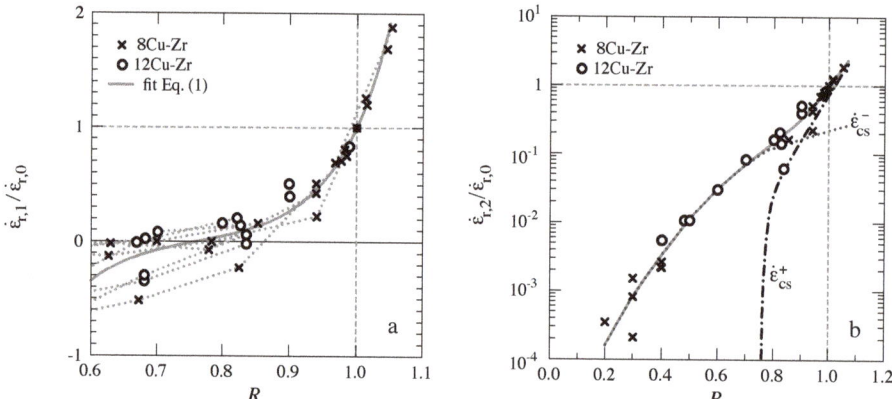

Figure 6. Normalized constant structure strain rate $\dot{\varepsilon}_r$ after qs deformation of 8/12Cu–Zr at $\sigma_{r,0} \approx 275$ MPa as function of relative creep load R: (**a**) $\dot{\varepsilon}_{r,1}$, grey dotted lines connect data from same test with stepwise load reduction, (**b**) $\dot{\varepsilon}_{r,2}$ (symbols connected by solid grey line); on log scale with estimates of $\dot{\varepsilon}_{cs}^-$ (dotted black) and $\dot{\varepsilon}_{cs}^+$ (dash-dotted black); see text.

4. Discussion

Our results for ufg Cu–Zr are qualitatively quite similar to the general behavior observed for crystalline materials after a perturbation of monotonic plastic flow by load changes. For small R-reductions deformation goes on at reduced rate in forward direction according to the applied stress and the material softens with strain in parallel to the recovery of the dislocation structure. For large R-reductions deformation first goes backward before it returns to positive direction again and then continues at decreasing rate. As mentioned in Equation (1), this rate decrease parallels that of recovery and therefore may be directly linked to dynamic recovery. This can be understood from the view that the strain rate term $\dot{e}^+$ leading to storage of dislocations disappears for small R so that the strain rate term $\dot{e}^-$ related with dynamic recovery dominates. These transient phenomena disappear while the new qs state corresponding to R is approached.

The two terms $\dot{e}^+$ and $\dot{e}^-$, corresponding to the cases 'dislocations in' and 'dislocations out' of Figure 1, have different kinetics. This difference should become apparent in those ranges of R where either $\dot{e}^+$ or $\dot{e}^-$ dominate. This is in line with the different R-dependences of the lines for $\dot{e}^+$ and $\dot{e}^-$ in Figure 6b. Milička [14–16] restricted his measurements to the R-range with $\dot{e}_{r,1} \geq 0$. In spite of this restriction, he discovered that a single mechanism of deformation obeying Equation (4) is not sufficient to describe the variation of $\dot{e}_{r,1}$ with R. So he proposed to express $\dot{e}_{r,1}$ as a sum of two terms [15,16]. This parallels the separation of $\dot{e}_{pl}$ into $\dot{e}^+$ and $\dot{e}^-$ in Equation (2).

4.1. Strain Related with Storage of Defects

From the preceding discussion we surmise that for $R \leq 0.7$ the rate $\dot{e}_{r,2}$ approximately equals $\dot{e}^-$. Extrapolating the $\dot{e}_{r,2}$-curve for $R < 0.7$ in Figure 6 yields $\dot{e}_{qs}^-$-values at $R = 1$ in the range of 10% and 30% of $\dot{e}_{r,0}$. In other words: the recovery–strain rate $\dot{e}_{qs}^-$ contributes about $(20 \pm 10)\%$ to the qs strain rate. $\dot{e}_{cs}^+$ follows as the difference of $\dot{e}_r$ and $\dot{e}_{cs}^-$ (Equation (2)). The stress exponent of this curve at $R = 1$ is $n_{cs}^+ = 17$ at $R = 1$. This is close to the estimate 21 derived from the theory of thermally activated glide (Equation A15). In view of the simplifications and assumptions involved, we conclude from this result that an interpretation of $\dot{e}_{cs}^+$ in terms of the classical theory of thermally activated glide over fixed repulsive obstacles in pure materials (e.g., forest dislocations) may be possible.

4.2. Strain Related with Recovery of Defects

We now turn attention to the recovery–strain rate $\dot{e}^-$. Figure 7a compares the recovery–strain rates $\dot{e}_{cs}^-$ at (approximately) constant structure from Figure 6b (dotted line) to the recovery–strain rate $\dot{e}_{qs}^-$ at qs structure (solid line) as function of stress σ. The latter is obtained from the qs strain rates $\dot{e}_{qs} \propto \sigma^6$ reported in the companion paper [28] under the assumption that the fraction $\dot{e}_{qs}^-/\dot{e}_{qs}$ in qs deformation equals ≈ 0.2 independent of stress. $\dot{e}_{cs}^-$ is larger than $\dot{e}_{qs}^-$. This can be qualitatively explained by the higher defect density and higher local stresses in the cs states inherited from the preceding deformation at the high stress $\sigma_{r,0} \approx 275$ MPa compared to the qs states established at lower stresses $\sigma < \sigma_{r,0}$. So far there is no accepted detailed model of dynamic recovery and its strain rate contribution $\dot{e}^-$. Strain contributions from recovery of individual dislocations stored at recovery sites, probably internal crystal boundaries (LABs, HABs), and from recovery of boundaries by migration need to be considered.

One may ask to which extent the recovery–strain rate gets reduced in the period of back flow before $\dot{e}_{r,2}$ is measured. It is clear that anelastic back flow relaxes internal stresses. Also, some fast recovery processes of the kind shown in Figure 1 will happen already during the period of net back flow and thereby reduce the density of recovery sites. This indicates that use of the term 'constant structure' for $\dot{e}_{cs}^-$ becomes increasingly problematic with declining R with regard to the dislocation structure and raises the question whether the constant structure assumption is wrong and anelastic back flow may even be lasting long enough to modify not only the internal stresses, but also allow the dislocation structure to evolve close to the new qs state at reduced stress. In this case $\dot{e}_{r,2}$ should become equal to the qs rate $\dot{e}_{qs}$ for low R and correspondingly low stresses. And this is in fact observed around 100 MPa, as Figure 7a shows. To answer the question we estimate a lower limit t'_{rec} of the time t_{rec} for full recovery into the new qs state. The estimate is based on the assumptions that (i) no dislocation generation takes place during the anelastic back flow even though the new qs state is based on dynamic equilibrium of generation and recovery and (ii) the maximal rate of dislocation recovery pertains throughout the back flow period even though the driving force for recovery must decrease. In the literature there is very little direct information on the evolution of the density ρ of dislocations during dynamic recovery. The reasons are that dynamic recovery is generally accompanied by dislocation glide of type $\dot{e}^+$ and that reliable observations can only be made if the dislocations can safely be pinned up to microscopic observation. A set of data was measured in [20] for the alloy Al–Zn where pinning is possible by precipitation of particles. The data were obtained in the qs state characterized by Equation (1). It was found that the measured dislocations recovery rates $\approx \dot{\rho}^-$ were in accord with Equation (1) when the dislocation generation rate is expressed as

$$\dot{\rho}^- \approx \dot{\rho}^+ = \frac{M f_\Lambda}{b} \frac{\dot{\varepsilon}^+}{\Lambda}. \qquad (5)$$

where Λ is proportional to the mean free path of dislocations and f_Λ is a numerical factor near 1. For a rough estimate we set $\Lambda = d_0$, $\dot{\varepsilon}_{r,0} = 10^{-4}\,\text{s}^{-1}$, $f_\Lambda = 1$. This yields the rate $\dot{\rho}_0^-$ of dynamic dislocation recovery just before the R-reduction as $2 \times 10^{-12}\,\text{m}^2\,\text{s}^{-1}$. The initial qs dislocation spacing is estimated as $\rho_{qs} = (b\,G/\sigma_{r,0})^2$ at $\sigma_{r,0} = 275\,\text{MPa}$. The solid line in Figure 4 shows the result for t'_{rec}. The data symbols represent the experimental data for the time period t_{back} where anelastic back flow occurs or cannot be excluded due to experimental inaccuracy. The result of this estimate is that in a large R-range the time period t_{back} available for recovery during back flow is smaller than the lower bound t'_{rec} of the time period t_{rec} of recovery needed to reach the new qs state of dislocation density. This corresponds to the observation made in situ on nanocrystalline Ni that recovery of X-line widths continues after the period of back flow [22]. So we conclude that $\dot{\varepsilon}_{cs}^-$ in Figure 6b mainly represents the recovery–strain rate due to $\dot{\rho}^-$, and not the qs strain rate resulting from $\dot{\rho}^+$-$\dot{\rho}^-$-balance (Equation (1)).

The results of the present work do not allow us to deduce details about the mechanism of recovery–strain. Cross slip [31] and climb [32] are generally being considered as rate-controlling mechanisms (compare [3]). Stress concentrations at boundaries by long-range internal stresses have been used in descriptions of kinetics with the composite model [33]. LABs in coarse-grained materials [11,21] and of HABs in nanocrystalline materials [34] are being discussed as sinks of dislocations as well as of boundaries themselves (via recombination during migration). Measurements on single-crystalline LiF have led to the conclusion that migration of LABs is responsible for most or even all of the observed recovery strain [21]. (A different situation is encountered in class II alloys like Al–Mg with viscous dislocation glide due to strong solute drag and spatially homogeneous distribution of recovery events [2,18,19,35]; here long-range stresses seem to play only little role.) The insensitivity of the relative recovery–strain contribution of recovery in the qs state to the boundary misorientation is intriguing. The observation that LABs are the major carriers of recovery strain in coarse-grained (cg) materials means that generation of recovery–strain by annihilation of single dislocations (Figure 1) cannot be used to explain the recovery strain in both LAB- and HAB-dominated structures. A boundary-mediated recovery mechanism, however, may be valid in both cases. It could mean that the rates of dynamic recovery vary with the HAB-content, but the basic mechanism involving free dislocations and boundaries is the same.

Better understanding of recovery–strain may be of profound value in technical application of strong materials under conditions of varying stress σ, e.g., in stress relaxation and cyclic deformation. The period of dominant recovery–strain rate $\dot{\varepsilon}^-$ after load reductions gives the unique chance to investigate the kinetics of dynamic recovery alone without influence of the storage strain rate $\dot{\varepsilon}^+$. One option is to perform *secondary* load change tests in this period after a primary large load reduction. Such secondary load changes have been started on Al [36] and recently continued on nc Ni [22,23]. The stress sensitivity in the period of $\dot{\varepsilon}_{qs}^-$-dominance was found to be much smaller than the qs stress sensitivity n_{qs}. Another option is to develop a *model* of dynamic recovery describing both the stress dependences of $\dot{\varepsilon}_{cs}^-(R)$ at constant structure and of $\dot{\varepsilon}_{qs}^-$ in the qs state.

An obvious question to be answered by a model is why the recovery–strain rate $\dot{\varepsilon}_{cs}^-$ at constant structure is so similar to the qs strain rate $\dot{\varepsilon}_{qs}$ (Figure 7a). At least part of the answer seems to lie in internal stresses [37]. It is probable that recovery processes are concentrated at relatively hard regions, in particular crystallite boundaries, where the local stress σ_h is enhanced relative to the applied stress by a local forward stress σ_f. In the following we apply the phenomenological approach used in [37] to the present case. Assume that $\dot{\varepsilon}^-$ varies with a power q of σ_h:

$$\dot{\varepsilon}_{cs}^- = f_0^- \dot{\varepsilon}_{r,0} \left(\frac{\sigma_h}{\sigma_{h,0}}\right)^q ; \qquad (6)$$

f_0^- connects $\dot{\varepsilon}_{cs}^-$ to the strain rate $\dot{\varepsilon}_{r,0}$ before the stress reduction; σ_h is the sum of applied stress $R\sigma_{r,0}$ after R-reduction and local internal forward stress $\sigma_f = f_{rel}\sigma_{f,0}$, i.e., $\sigma_h = R\sigma_{r,0} + f_{rel}\sigma_{f,0}$; here $\sigma_{f,0}$ is the forward stress at $R = 1$ before the load reduction and f_{rel} describes the relaxation of the internal forward stress during anelastic back flow before $\dot{\varepsilon}_{cs}^-$ is measured; $\sigma_{h,0}$ is the starting value of σ_h at $R = 1$ and $f_{rel} = 1$ before the R-change. To give an example, f_0^- is set to 0.22, the exponent in Equation (6) is chosen as $q = 7$, and $\sigma_{f,0}$ is assumed to be $1.5\sigma_{r,0}$. With these choices Figure 7a shows $\dot{\varepsilon}_{cs}^-$ as function of R for two cases. The first case $f_{rel} = 1$, i.e. no relaxation of internal forward stress during anelastic back flow, yields $\dot{\varepsilon}_{cs}^-$-values lying distinctly higher than the measured ones, but is unrealistic. In the second case f_{rel} is assumed to decreases with decreasing R as shown in Figure 7b. The thick dotted curve in Figure 7a represents the result for $\dot{\varepsilon}_{cs}^-$. It was made to perfectly match the measured $\dot{\varepsilon}_{cs}^-$-curve from Figure 6b. For comparison, the line for $\dot{\varepsilon}_{qs}^-$ shows the recovery–strain rate expected in the qs state, if the ratio f_0^- is independent of stress as suggested by numerous results obtained for cg materials. It is seen that although the relaxation of the internal stresses during back flow may be significant, it always keeps the recovery–strain rate generated from the dense defect structure at high stress $\sigma_{r,0}$ above the qs strain rate expected at the lower stresses acting at $R < 1$ where the qs defect density is much lower. That makes sense. The preceding exercise shows that the measured constant structure recovery–strain rates can be understood on the basis of internal forward stresses of some kind acting at the recovery sites. It must, however, naturally be expected that the decrease of the volume density of recovery sites during back flow also contributes to the decline of $\dot{\varepsilon}_{cs}^-$, qualitatively marked by the downward pointing arrows in Figure 7.

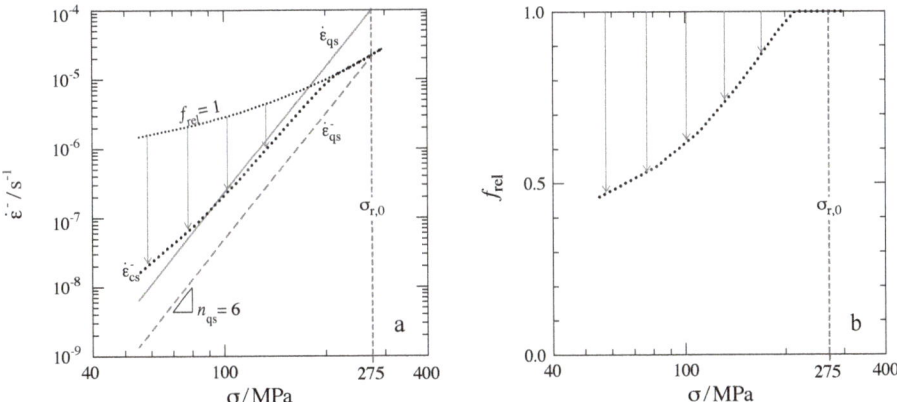

Figure 7. (a) Recovery–strain rate $\dot{\varepsilon}_{cs}^-$ at constant structure after R-change from Figure 6b (black dotted) compared to qs strain rate $\dot{\varepsilon}_{qs}$ (grey solid) and recovery–strain rate $\dot{\varepsilon}_{qs}^-$ in the qs state (grey dashed), (b) anelastic relaxation factor f_{rel} as function of $\sigma \approx R\sigma_{r,0}$ required to model $\dot{\varepsilon}_{cs}^-$ from (a) with Equation (6).

4.3. Comparison of Stress Dependences of $\dot{\varepsilon}^+$ and $\dot{\varepsilon}^-$ at Constant Structure

One problem with measuring the recovery–strain rate $\dot{\varepsilon}^-$ is that its separation from $\dot{\varepsilon}^+$ in load/stress change tests is not trivial and sometimes impossible. The separation is easy and accurate if the inflection point in the semilogarithmic $\dot{\varepsilon}_{r,2}$-curve (Figure 6b) is well pronounced. This depends strongly on the slope of this curve at $R = 1$. This slope is mostly given by the stress exponent n_{cs}^+ of $\dot{\varepsilon}^+$ (see Equation A15), i.e., the rate associated with generation of defects leading to work hardening. According to the estimate of Equation A14 n_{cs}^+ decreases inversely proportional to the temperature T.

On the other hand, the slope of the $\dot{\varepsilon}^-$-curve due to recovery is rather insensitive to T. Therefore, the separation of $\dot{\varepsilon}^-$ becomes increasingly problematic when T increases. Solid solution strengthening leads to further reduction of n_{cs}^+ and the inflection point in the semilogarithmic $\dot{\varepsilon}_{r,2}$-curve (Figure 6b) may disappear completely (e.g., in Al–5Mg [14,19] and Fe–Si [15]). Then the separation of $\dot{\varepsilon}^+$ and $\dot{\varepsilon}^-$ may be based on the fact that $\dot{\varepsilon}^+$ is driven by a thermal stress component *lower* than the applied stress, whereas $\dot{\varepsilon}^-$ is driven by a local stress that is *enhanced* by the interaction of the recovering defects; this is an open task.

5. Summary

- In ufg Cu–Zr at 0.5 T_m recovery–strain ε^- connected with dynamic recovery of strain-induced crystal defects was found in tests with perturbation of the quasi-stationary (qs) state by load reductions. ε^- adds to the strain ε^+ connected with dislocation generation and storage.
- The stress dependence of $\dot{\varepsilon}^+$ yields an activation volume consistent with the classical theory of thermally activated glide.
- The recovery–strain rate $\dot{\varepsilon}^-$ contributes 10% to 30% to the quasi-stationary strain rate $\dot{\varepsilon}_{qs}$. This fraction for ufg Cu–Zr with high volume fraction of HABs is similar to the one commonly reported for cg materials with high volume fraction of LABs. That could mean that boundaries play qualitatively similar roles in mediating dynamic recovery independent of their misorientation.
- The values of $\dot{\varepsilon}_{cs}^-$ (at constant structure) and $\dot{\varepsilon}_{qs}$ (at quasi-stationary structure) are relatively similar for large load reductions, even though the microstructures, in particular the dislocation structures, should differ significantly. This becomes understandable, if promotion of recovery by internal forward stresses is taken into account.
- Combining the rates of recovery–strain in the qs state and after perturbation of monotonic flow seems promising to better understand the mechanism of dynamic recovery of crystal defects, limiting the deformation strength under monotonic as well as cyclic loading conditions.

Author Contributions: Conceptualization, W.B. and P.K.; methodology, V.S., J.D., P.K. and P.E.; software, W.B. and P.E.; validation, P.K., J.D.; formal analysis, W.B.; investigation, V.S., J.D., P.K.; resources, J.D.; data curation, J.D., P.K. and W.B.; writing–original draft preparation, W.B.; writing–review and editing, W.B., P.K. and P.E.; visualization, W.B; J.D., P.K., W.B. and P.E.; supervision, W.B. and V.S.

Conflicts of Interest: The authors declare no conflict of interest.

Abbreviations

The following abbreviations are used in this manuscript:

qs	quasi-stationary
ECAP	equal channel angular pressing
cg	coarse-grained
ufg	ultrafine-grained
LAB	low-angle boundary
HAB	high-angle boundary

Appendix A. Determination of Inelastic Strain

The load F corresponding to an engineering stress

$$\sigma_{eng} = F/S_0 \tag{A1}$$

was varied in steps. Figure A1a shows an example. Assuming volume constancy, the cross section varies with the gauge length

$$l = l_0 + \Delta l, \tag{A2}$$

where Δl is the measured length change, as

$$S = S_0 l_0 / l = S_0 \exp(-\epsilon_{tot}), \qquad \epsilon_{tot} = \ln(l/l_0) \tag{A3}$$

where ϵ_{tot} is the total "true" strain. Figure A1b shows the variation of ϵ_{tot} with time t corresponding to Figure A1a. The ϵ_{tot}-steps in Figure A1b result from the changes of the elastic strain related with the changes of F. To eliminate these steps the elastic strain must be estimated. This was done in the following straightforward manner. The elastic strain is composed from two components:

$$\epsilon_{el} = \epsilon_{el,Cu} + \epsilon_{mach}. \tag{A4}$$

$\epsilon_{el,Cu}$ is the elastic strain of the gauge length l of the specimen described by:

$$\epsilon_{el,Cu} = \sigma / E \tag{A5}$$

with

$$\sigma = F_c / S \approx \sigma_{eng} \exp(\epsilon_{tot}) \tag{A6}$$

as "true" stress acting in the gauge length and $E \approx 9 \times 10^4$ MPa as elastic tensile modulus (Young's modulus) of Cu. ϵ_{mach} is the elastic strain

$$\epsilon_{mach} = \Delta l_{mach} / l \tag{A7}$$

resulting from all parts of specimen and machine entering the measured length change outside the gauge length l. The unknown elastic machine length change was determined in an iterative manner so that the elastic steps in the ϵ_{tot}–t plots like Figure A1b were optimally suppressed.

An analytical formulation with a power law:

$$\Delta l_{mach} / \text{mm} \approx c_1 \, (F_c / \text{N})^{c_2} - c_3. \qquad 0.001 < c_3 < 0.006 \tag{A8}$$

with $c_1 = 2.23 \times 10^{-4}$, $c_2 = 0.74$ and a constant c_3 turned out to be comfortable and sufficiently exact. The approximate inelastic strain then follows as:

$$\epsilon_{inel} = \epsilon_{tot} - \epsilon_{el}. \tag{A9}$$

Individual choice of c_3 for each test proved reasonable to compensate systematic errors of the Δl-signal near $F = 0$ before the motions of specimen and strain gages become uniaxial. In a final step the stress was corrected by changing Equation (A6) to

$$\sigma \approx F_c / S = \sigma_{eng} \exp(\epsilon_{inel}). \tag{A10}$$

This has only marginal influence on the results. Figure A1c shows that the elastic steps from Figure A1b have virtually disappeared. Some gaps in the curves are caused by data acquisition problems. The test includes a small stress increase at $t \approx 300$ s followed by stepwise unloading within less than 30 s. It is seen how the (inelastic) strain ϵ_{inel} continues to increase till 307 s and then starts to decrease. This decrease is called anelastic, because it is reversible on a macroscopic level. The elimination of the elastic strain helps to visualize the anelastic response that is less pronounced than the elastic one (also in comparison to the elastic response of the specimen). Equation (A7) may cause an elastic overcorrection at stresses below 100 MPa. However, this is irrelevant for the inelastic strain rates in the periods of relatively constant load, where the major elastic strain component resulting from Δl_{mach} remains constant.

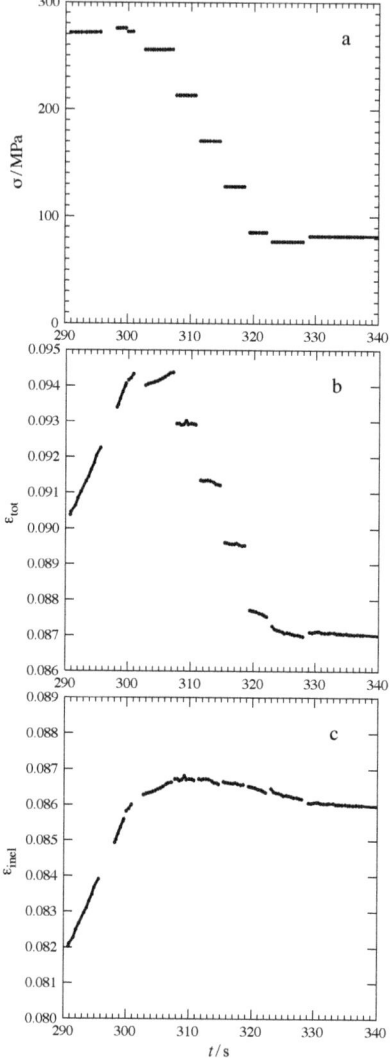

Figure A1. (a) Stress σ, (b) total strain ϵ_{tot} with elastic strains from machine and specimen, (c) inelastic strain ϵ_{inel} versus time t in load change test on 8Cu–Zr at 673 K.

Appendix B. Activation Volume of Dislocation Glide

Glide in the course of expansion of dislocation loops bounding the slipped areas causes an inelastic strain rate $\dot{\epsilon}^+$. It is driven by the resolved shear stress σ/M, where M is the geometrical factor of conversion from normal to shear stress and strain (for untextured face-centered polycrystals: Taylor factor = 3.06), k_B is the Boltzmann constant, and is supported by thermally activated overcoming of thermal obstacles. The operational activation volume is defined by

$$V_{op}^+ = k_B T \frac{d\ln \dot{\epsilon}^+}{d\sigma/M} \qquad (A11)$$

To get a rough estimate of V_{gl}^{op} we tentatively use the classical model of thermally activated glide through a field of point-like repulsive obstacles. According to this model the activation volume is

$$V^+ = b \lambda_{gl} \Delta x_{gl}. \qquad (A12)$$

where λ_{gl} and Δx_{gl} are obstacle spacing and width, respectively. Equation (A12) holds under the condition that the microstructure including the internal stresses remains constant in the change test. If

- λ_{gl} is set equal to the expected spacing of free dislocations, bG/σ, and
- Δx_{gl} is approximated by b,

V^+ becomes a simple function of stress:

$$V^+ \approx b^3 G/\sigma. \qquad (A13)$$

By approximating V_{op}^+ in Equation (A11) by V^+ from Equation (A13) and using the mathematical identity $d\sigma = \sigma d\ln\sigma$ one arrives at a simple estimate

$$n_{cs,est}^+ \equiv \frac{b^3 G}{M k_B T}. \qquad (A14)$$

of the stress exponent of $\dot{\epsilon}^+$ at constant structure:

$$n_{cs}^+ = \frac{\partial \ln \dot{\epsilon}^+}{\partial \ln \sigma}. \qquad (A15)$$

(Meanwhile it has become customary to neglect the condition of constant structure; this leads to a mix-up with the qs rate sensitivity [16,38].) The estimate $n_{cs,est}^+$ is independent of σ and inversely proportional to temperature T for a given material.

citeyearref-journal-3b, p.475). This produces: Wong (1999, p. 328; 2000, p. 475)

References

1. Sherby, O.D.; Burke, P.M. Mechanical Behaviour of Crystalline Solids at Elevated Temperature. *Progr. Mater. Sci.* **1968**, *13*, 323–390, doi:10.1016/0079-6425(68)90024-8. [CrossRef]
2. Takeuchi, S.; Argon, A. Steady-state creep of alloys due to viscous motion of dislocations. *Acta Metall.* **1976**, *24*, 883–889, [CrossRef]
3. Čadek, J. *Creep in Metallic Materials*; Elsevier: Amsterdam, The Netherlands, 1988.
4. Blum, W.; Nix, W.D. Strain associated with recovery. *Res. Mech. Lett.* **1981**, *1*, 235–240.
5. Blum, W. On the Evolution of the Dislocation Structure during Work Hardening and Creep. *Scr. Metall.* **1984**, *18*, 1383–1388, doi:10.1016/0036-9748(84)90370-3. [CrossRef]
6. Mohamed, F.A.; Langdon, T.G. The transition from dislocation climb to viscous glide in creep of solid solution alloys. *Acta Metall.* **1974**, *22*, 779–788. [CrossRef]
7. Lindroos, V.; Miekk-oja, H. The structure and formation of dislocation networks in aluminium-magnesium alloys. *Phil. Mag.* **1967**, *16*, 593–610. [CrossRef]
8. Blum, W. Dislocation Models of Plastic Deformation of Metals at Elevated Temperatures. *Z. Metallkde.* **1977**, *68*, 484–492.
9. Caillard, D. In situ creep experiments in weak beam conditions, in Al at intermediate temperature, Interaction of dislocations with subboundaries. *Acta Metall.* **1984**, *32*, 1483–1491. [CrossRef]

10. Petegem, S.V.; Brandstetter, S.; Schmitt, B.; Swygenhoven, H.V. Creep in nanocrystalline Ni during X-ray diffraction. *Scr. Mater.* **2009**, *60*, 297–300, doi:10.1016/j.scriptamat.2008.10.034. [CrossRef]
11. Exell, S.F.; Warrington, D.H. Sub-grain Boundary Migration in Aluminum. *Philos. Mag. A* **1972**, *26*, 1121–1136. [CrossRef]
12. Gorkaya, T.; Molodov, D.A.; Gottstein, G. Stress-driven migration of symmetrical ⟨100⟩ tilt grain boundaries in Al bicrystals. *Acta Mater.* **2009**, *57*, 5396–5405. [CrossRef]
13. Humphreys, F.J.; Hatherly, M. *Recrystallization and Related Annealing Phenomena*; Elsevier Science: Oxford, UK, 1995.
14. Milička, K. Constant structure creep in metals after stress reduction in steady state stage. *Acta Metall. Mater.* **1993**, *41*, 1163–1172. [CrossRef]
15. Milička, K. Constant structure experiments in high temperature primary creep of some metallic materials. *Metall. Mater.* **1994**, *42*, 4189–4199. [CrossRef]
16. Milička, K. Constant structure creep experiments on aluminium. *Kov. Mater.* **2011**, *49*, 307–318. [CrossRef]
17. Nix, W.D.; Ilschner, B. Mechanisms Controlling Creep of Single Phase Metals and Alloys. In *Proceedings of the 5th International Conference, Aachen, Germany, 27–31 August 1979*; Haasen, P., Gerold, V., Kostorz, G., Eds.; Pergamon Press: Oxford, UK, 1980; pp. 1503–1530.
18. Weckert, E.; Blum, W. Transient Creep of an Al-5 at%Mg Solid Solution. In *Draft of paper in Proc. 7th Int. Conf. on the strength of Metals and Alloys (ICSMA7), Montreal, 12–16 August 1985*; McQueen, H.J., Bailon, J.P., Dickson, J.I., Jonas, J.J., Akben, M.G., Eds.; Pergamon Press: London, UK, 1985; pp. 773–778.
19. Blum, W.; Weckert, E. On the Interpretation of the 'Internal Stress' determined from Dip Tests during Creep of Al-5 at% Mg. *Mater. Sci. Eng.* **1987**, *23*, 145–158. [CrossRef]
20. Hausselt, J.; Blum, W. Dynamic recovery during and after steady state deformation of Al-11wt%Zn. *Acta Metall.* **1976**, *24*, 1027–1039.10.1016/0001-6160(76)90133-4. [CrossRef]
21. Müller, W.; Biberger, M.; Blum, W. Subgrain Boundary Migration during Creep of LiF, III. Stress Reduction Experiments. *Philos. Mag. A* **1992**, *66*, 717–728. [CrossRef]
22. Sun, Z.; Van Petegem, S.; Cervellino, A.; Durst, K.; Blum, W.; Van Swygenhoven, H. Dynamic recovery in nanocrystalline Ni. *Acta Mater.* **2015**, *91*, 91–100.10.1016/j.actamat.2015.03.033. [CrossRef]
23. Sun, Z.; Van Petegem, S.; Cervellino, A.; Blum, W.; Van Swygenhoven, H. Grain size and alloying effects on dynamic recovery in nanocrystalline metals. *Acta Mater.* **2016**, *119*, 104–114, doi:10.1016/j.actamat.2016.08.019. [CrossRef]
24. Hasegawa, T.; Yakou, T.; Kocks, U. Length changes and stress effects during recovery of deformed aluminum. *Acta Metall.* **1982**, *30*, 235–243. [CrossRef]
25. Hasegawa, T.; Yakou, T.; Kocks, U.F. Forward and Reverse Rearrangements of Dislocations in Tangled walls. *Mater. Sci. Eng.* **1986**, *81*, 189–199. [CrossRef]
26. Kapoor, R.; Li, Y.J.; Wang, J.T.; Blum, W. Creep transients during stress changes in ultrafine-grained copper. *Scr. Mater.* **2006**, *54*, 1803–1807.10.1016/j.scriptamat.2006.01.032. [CrossRef]
27. Blum, W. A new way to visualize back strain and recovery strain after stress reductions. *ResearchGate* **2019**.10.13140/RG.2.2.18688.69127/2. [CrossRef]
28. Dvořák, J.; Blum, W.; Král, P.; Sklenička, V. Quasi-stationary strength and strain softening of Cu–Zr after predeformation by ECAP. *Preprints* **2019**, 2019090073. [CrossRef]
29. Frost, H.J.; Ashby, M.F. *Deformation-Mechanism Maps*; Pergamon Press: Oxford, UK, 1982.
30. Eisenlohr, P. SmooMuDS: Flexible Smoothing of Multi-Valued Data Series. 2006. Available online: http://www.gmp.ww.uni-erlangen.de/SmooMuDS_engl.php (accessed on 25 November 2019). doi:10.13140/2.1.3636.2566. [CrossRef]
31. Kocks, U.; Mecking, H. Physics and phenomenology of strain hardening: the FCC case. *Progr. Mater. Sci.* **2003**, *48*, 171–273.10.1016/S0079-6425(02)00003-8. [CrossRef]
32. Nes, E. Modelling work hardening and stress saturation in FCC metals. *Progr. Mater. Sci.* **1998**, *41*, 129–193.10.1016/S0079-6425(97)00032-7. [CrossRef]
33. Vogler, S.; Blum, W. Micromechanical Modelling of Creep in Terms of the Composite Model. In *Creep and Fracture of Engineering Materials and Structures*; Wilshire, B., Evans, R., Eds.; The Institute of Metals: London, UK, 1990; pp. 65–79.
34. Dupraz, M.; Sun, Z.; Brandl, C.; Van Swygenhoven, H. Dislocation interactions at reduced strain rates in atomistic simulations of nanocrystalline Al. *Acta Mater.* **2018**, *144*, 68–79. [CrossRef]

35. Kassner, M.; Pérez-Prado, M.T. Five-Power-Law Creep in Single Phase Metals and Alloys. *Progr. Mater. Sci.* **2000**, *45*, 1–102.10.1016/S0079-6425(99)00006-7. [CrossRef]
36. Blum, W.; Rosen, A.; Cegielska, A.; Martin, J.L. Two mechanisms of dislocation motion during creep. *Acta Metall.* **1989**, *37*, 2439–2453. [CrossRef]
37. Mekala, S.; Eisenlohr, P.; Blum, W. Control of dynamic recovery and strength by subgrain boundaries—Insights from stress-change tests on CaF_2 single crystals. *Philos. Mag.* **2011**, *91*, 908–931.10.1080/14786435.2010.535324. [CrossRef]
38. Blum, W. Discussion: Activation volumes of plastic deformation of crystals. *Scr. Mater.* **2018**, *146*, 27–30.10.1016/j.scriptamat.2017.10.029. [CrossRef]

© 2019 by the authors. Licensee MDPI, Basel, Switzerland. This article is an open access article distributed under the terms and conditions of the Creative Commons Attribution (CC BY) license (http://creativecommons.org/licenses/by/4.0/).

Article

Quasi-Stationary Strength of ECAP-Processed Cu-Zr at 0.5 T_m

Wolfgang Blum [1,*], Jiří Dvořák [2], Petr Král [2], Philip Eisenlohr [3] and Vaclav Sklenička [2]

1. Department of Materials Science, Institute I, University of Erlangen-Nuremberg, Martensstr. 5, D-91058 Erlangen, Germany
2. Institute of Physics of Materials, Czech Academy of Sciences, Žižkova 22, CZ-616 62 Brno, Czech Republic; dvorak@ipm.cz (J.D.); pkral@ipm.cz (P.K.); sklen@ipm.cz (V.S.)
3. Chemical Engineering and Materials Science, Michigan State University, East Lansing, MI 48824, USA; eisenlohr@egr.msu.edu
* Correspondence: wolfgang.blum@fau.de

Received: 5 September 2019; Accepted: 21 October 2019; Published: 26 October 2019

Abstract: The influence of the grain structure on the tensile deformation strength is studied for precipitation-strengthened Cu-0.2%Zr at 673 K. Subgrains and grains are formed by equal channel angular pressing (ECAP) and annealing. The fraction of high-angle boundaries increases with prestrain. Subgrains and grains coarsen during deformation. This leads to softening in the quasi-stationary state. The initial quasi-stationary state of severely predeformed, ultrafine-grained material exhibits relatively high rate-sensitivity at relatively high stresses. This is interpreted as a result of the stress dependences of the quasi-stationary subgrain size and the volume fraction of subgrain-free grains.

Keywords: Cu-Zr; ECAP; deformation; quasi-stationary; subgrains; grains; coarsening

1. Introduction

Deformation of single crystals and large-grained polycrystals leads to generation of dislocations and subgrains with low-angle boundaries (LABs). Dynamic recovery counteracts the defect generation. The state at which the rate $\dot\rho^+$ of increase of density ρ of free dislocations approximately equals the rate $\dot\rho^-$ of ρ-decrease,

$$\dot\rho^+ \approx \dot\rho^-, \qquad (1)$$

is usually called steady state, stationary state or saturation stage (of flow stress) in the literature. We prefer the term quasi-stationary (qs) state [1], because slow changes of microstructure parameters (boundaries [2,3], particles, ...) lead to a drift of the quasi-equilibrium of the qs state that in turn causes slow changes of the qs deformation strength. By definition, there is a spectrum of qs states, depending on the microstructural parameters other than ρ. The qs states are important as upper limits of deformation strength at given structure of grains and particles.

The present work deals with the strength of a Cu-Zr alloy that is stabilized by particles and processed by equal channel angular pressing (ECAP). It is well known that ECAP processing leads into a qs state with refined grains and saturated hardness (compare [4–7]). This holds also for Cu-Zr [8–10]. When the ECAP-processed material is subjected to creep at elevated temperature of 673 K, new qs states are developed. The creep ductility is significantly enhanced compared to the coarse-grained (cg) alloy while the creep rate remains fairly low and has relatively low stress sensitivity. In the following section, we treat the evolution of deformation strength and microstructure of ECAP-processed Cu-Zr at 673 K in detail on updated data basis with the goal to better understand how grain refinement influences the qs deformation strength. We discuss the proposition that the qs strength of ultrafine-grained (ufg)

Cu-Zr significantly depends on the stress-dependent qs fraction of subgrain-free grains. There are indications that the variation of this fraction with stress may explain the relatively low stress sensitivity (high rate sensitivity) of the qs strength of grain-refined ufg Cu-Zr.

2. Experimental Details

Processing, microstructure, and creep testing have been described before [10]. The starting material was (cg) Cu-0.2 mass%Zr. Its material parameters are approximated by those of pure Cu provided in the data compilation of Frost and Ashby [11]: Burgers vector $b = 2.56 \times 10^{-10}$ m, elastic shear modulus $G = 3.58 \times 10^4$ MPa, melting point $T_m = 1356$ K. The cg material was homogenized for 24 h at 1073 K and hot rolled. Billets of this material were solution treated at 1233 K for 1 h to give an initial grain size of 3.5×10^{-4} m. The billets were predeformed by p passes of equal channel angular pressing (ECAP, Figure 1) at room temperature on route B_C to refine the grains. The predeformed material is called pCu-Zr. Due to particle hardening the grain size of the material is stable against static annealing at temperatures exceeding the test temperature 673 K [12]. The thermal treatment consisted of storage of the predeformed material at room temperature for up to several years followed by heating to and holding at T for a few hours before deformation.

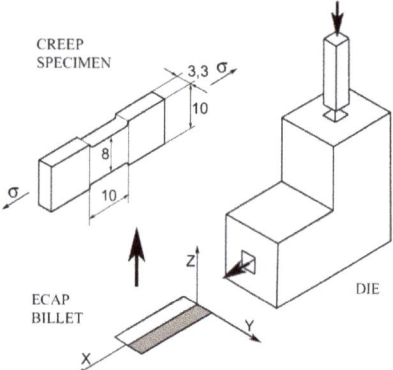

Figure 1. From ECAP billet to tensile creep specimen (numbers: lengths/mm).

Flat tensile creep specimens with gauge length l_0 and cross section S_0 were prepared as illustrated in Figure 1. Deformation in creep mode occurred in tension at temperature $T = 673$ K $= 0.5\,T_m$ upon applying a load F, corresponding to an engineering stress $\sigma_{eng} = F/S_0$. The change Δl of the length was measured using LVDTs (from Hottinger Baldwin Messtechnik GmbH, Darmstadt, Germany) attached to the tensile rods near the specimen. The "true" strain ϵ results as $\ln((l_0 + \Delta l)/l_0)$ (Elastic contribution to Δl are neglected here because it is relatively small and nearly constant at constant F so that the elastic contribution $\dot\epsilon_{el}$ to the measured rate $\dot\epsilon$ is generally negligible). Due to volume constancy and under uniform deformation of the gauge length $l = l_0 + \Delta l$, the cross section S varies as $S = S_0 l_0/l = S_0 \exp(-\epsilon)$ and the "true" stress σ, given by F/S, increases with strain as

$$\sigma = \sigma_{eng} \exp(\epsilon). \tag{2}$$

Observations of the grain structure were made by electron microscopy. Orientation maps derived from electron backscatter diffraction (EBSD, device from Oxford Instruments, High Wycombe, United Kingdom) were used to determine the grain size d as mean value of the spacings d_i, $i = 1, 2, 3, \ldots$, of HABs between crystallites with misorientations $\geq 15°$ determined along test lines. Transmission electron microscopy (TEM) images (microscope from JEOL Ltd., Tokyo, Japan) were used to analogously determine the subgrain size w as mean value of the spacings w_i of boundaries of any kind.

3. Results

3.1. Influence of Predeformation

With increasing predeformation p the deformation strength of pCu-Zr saturates, because the areal fraction f_{HAB} of high-angle boundaries (HABs) increases up to a saturation level [9] where a qs state is established due to dynamic equilibrium of storage and dynamic recovery of dislocations. This statement is consistent with the observation that the Vickers hardness saturates with increasing predeformation p at $HV \approx 180$ [9]; using the formula $\sigma/\mathrm{MPa} \approx 3\,HV$, this corresponds to a saturation flow stress of 540 MPa in ECAP. An increase of f_{HAB} enhances storage as well as recovery of dislocations because HABs are stronger dislocation obstacles and better dislocation sinks than LABs. The net effect on deformation strength depends on whether storage or recovery are the dominant factors (compare, e.g., [13]).

Figure 2 shows exemplary curves of the evolution of deformation resistance with strain at constant engineering stress $\sigma_{eng} = 150$ MPa for different amounts p of ECAP predeformation. Deformation starts with application of the load F, and continues in creep mode when the creep stress σ_{eng} has been reached. The starting phase of the tests is connected with work hardening as is visible from the sharp decrease of strain rate $\dot{\epsilon}$ with ϵ. After strains of 0.02 to 0.05 the rate $\dot{\epsilon}$ increases. A trivial reason for $\dot{\epsilon}$-increase lies in the increase of σ with ϵ (Equation (2)) causing the creep rate to increase in the qs state as $\dot{\epsilon} \propto \sigma^{n_{qs}}$ where n_{qs} is the qs value of the stress exponent. The dotted line in Figure 2 shows that for severely predeformed Cu-Zr with $p \geq 8$ practically all of the $\dot{\epsilon}$-increase can be attributed to the increase of σ when $n_{qs} = 6$ (as reported before [9]). This means that at $\sigma_{eng} = 150$ MPa

- the specimens of 8/12Cu-Zr begin to deform in their qs state right from low strains in the order of 0.02, and
- changes of boundary and particle structures are negligible.

At about half the fracture strain the $\dot{\epsilon}$-increase accelerates. This indicates formation of local stress concentrations by external and internal necking during the fracture process.

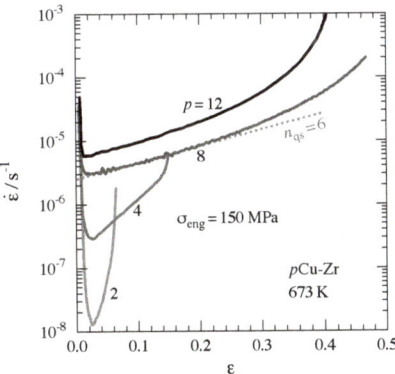

Figure 2. $\dot{\epsilon}$ versus ϵ for pCu-Zr at 673 K, $\sigma_0 = 150$ MPa in tension at constant load; dotted line: increase of $\dot{\epsilon}$ expected from increase of σ for $n_{qs} = 6$.

The less strongly predeformed specimens 4Cu-Zr and 2Cu-Zr exhibit a distinctly steeper increase of $\dot{\epsilon}$ with ϵ. To find the reason for that, it is useful to display the creep process in the qs region near the minimal creep rate, i.e., the creep between the work hardening period and the fracture period, in the

$\log \dot\varepsilon - \log \sigma$-field (Figure 3). According to Equation (2), $\log \sigma$ increases linearly with ε. The scale-bars in Figure 3 show the extension of a strain interval $\Delta\varepsilon$ of 0.5. Figure 3e confirms that most of the curves for 8Cu-Zr have slopes near $n_{qs} = 6$ in the qs range above 150 MPa and follow a common trend given by the dashed black line. For 12Cu-Zr this behavior is even more perfectly pronounced, indicating better homogeneity of the initial microstructure.

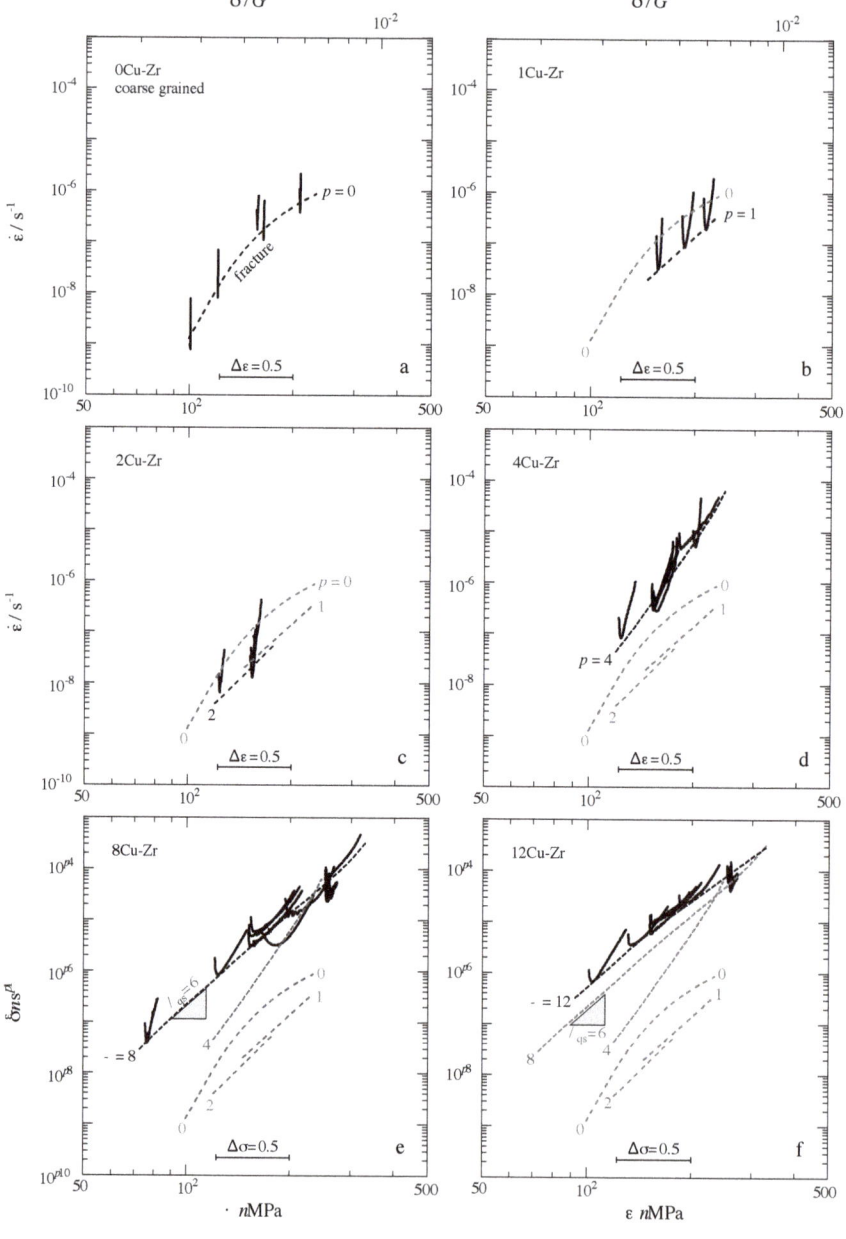

Figure 3. $\dot\varepsilon$ versus σ for pCu-Zr at 673 K in tension at constant load.

For lower stresses $\sigma < 150$ MPa and for pCu-Zr with $p \leq 4$ the slopes in Figure 3 are distinctly higher and different curves do not merge. One reason is tensile fracture. Figure 3a shows that fracture occurs in cg Cu-Zr (i.e., pCu-Zr with $p = 0$), before a qs region has been developed. This means that the qs deformation strength cannot be measured in tension. Thus, the experimental values of the minimal creep rate are determined by fracture, and therefore constitute only upper bounds of the qs deformation strength. The same holds for predeformed pCu-Zr with $p = 1$ to 4. (To determine the qs deformation strength, compression tests would be required in these cases in order to delay or suppress fracture). For ufg 8/12Cu-Zr the strains are too large to explain the non-merging of the curves at low σ by fracture processes. Therefore this effect must be due to softening during creep. In Section 3.3 we provide an explanation of the softening in terms of dynamic coarsening of the microstructure.

In summary, one sees from Figure 3 that the qs deformation strength of Cu-Zr decreases at the test temperature $0.5\,T_m$ when the material is severely predeformed, the grains are refined, and the areal fraction f_{HAB} of boundaries with high-angle character increases during creep.

3.2. Microstructure of 8Cu-Zr

3.2.1. Before Creep

Heating to and holding at test temperature before creep took a few hours. Figures 4 and 5a illustrate the microstructure of 8Cu-Zr after annealing for a few hours at the test temperature. So these microstructures are representative for the microstructure at start of creep. From a micrograph, such as Figure 4, the initial grain size d_0 of 8Cu-Zr was determined as 4.7×10^{-7} m. From the 8Cu-Zr sample of Figure 5a the initial subgrain size w_0 was determined as 1.5×10^{-7} m. Noting that these are local values that may vary from place to place and from specimen to specimen, the agreement with previously published data [10] and with data from the literature for severely predeformed ufg Cu-Zr [12,14,15] data is considered to be satisfactory. From the empirical formula

$$w_{qs}^{cg} \approx 14\,b\,G/\sigma \tag{3}$$

for cg Cu [16] and the estimate 540 MPa for the saturation strength in ECAP given above one gets a qs subgrain size of 2.4×10^{-7} m. In view of the uncertainty of the empirical factor 14 entering Equation (3), the agreement with the measured w_0 is considered satisfactory, too. This means that w_0 may be interpreted as the qs value of the subgrain size in ufg 8Cu-Zr before creep, that is only marginally influenced by the heating to test temperature.

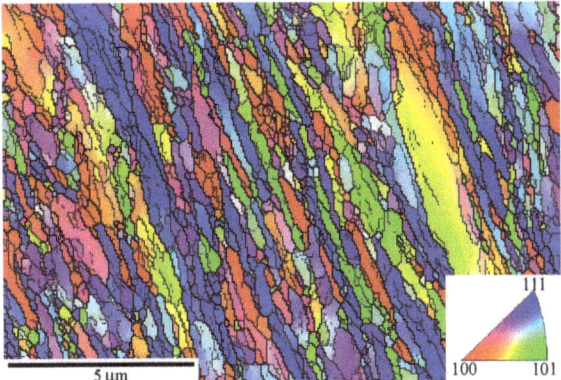

Figure 4. Grain structure map of 8Cu-Zr after 10 h of annealing at 673 K; thin and thick lines: traces of boundaries with misorientation $<$ and $> 15°$, respectively. Insert: color code for crystallographic grain normals in standard triangle.

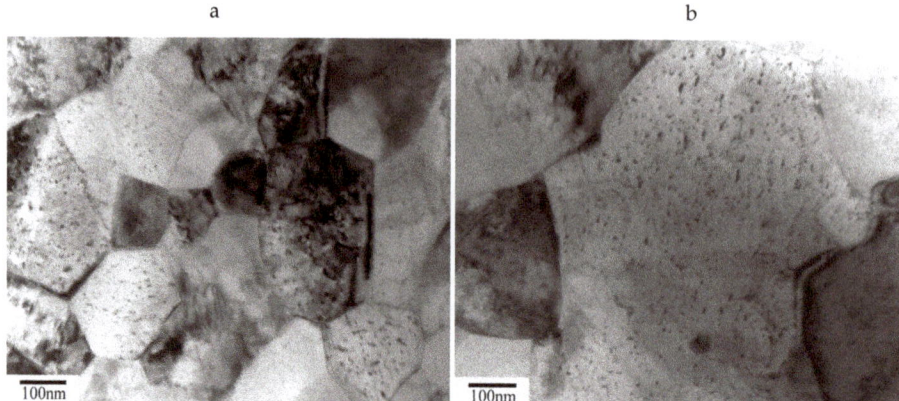

Figure 5. TEM of subgrains and Cu$_5$Zr-precipitates in 8Cu-Zr after (a) annealing for 8 h at 673 K, (b) creep at 673 K, $\sigma_{eng} = 150$ MPa to a fracture strain of ≈ 0.5 (compare curve for $p = 8$ in Figure 2).

The Zr containing particles are essential in restricting grain coarsening [10,12,14]. They are present in homogeneous distribution after SPD ([14] and Figure 5). Figure 6 shows electron diffraction patterns of (a) the Cu matrix with particles and (b) a large particle of composition Cu$_5$Zr. A rough estimate based on Equations (7)–(13) in [17] (using a mean particle radius of 4 nm, number of particles per area of TEM foil equal to 10, typical foil thickness of 200 nm) yields an Orowan stress for particle hardening of about 150 MPa at room temperature. However, at the test temperature the tiny particles are easily overcome by local and general climb of dislocations so that the particle hardening term is greatly reduced [17]. In the following, we neglect the direct influence of particles on the stress sensitivity of the deformation strength.

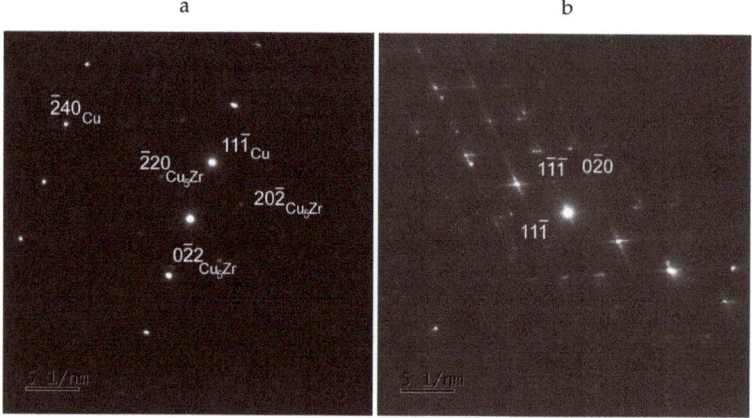

Figure 6. Diffraction patterns of (a) Cu matrix (strong reflections) with Cu$_5$Zr particles (weak reflections) and (b) a large Cu$_5$Zr particle.

The spacing plot of Figure 7 shows w_0 and d_0 together with the qs spacings $\approx bG/\sigma$ of dislocations and w_{qs}^{cg} of boundaries in cg Cu. We assume that the qs spacings for cg Cu are valid also for Cu-Zr. It is interesting to note that at low stresses the (dotted) qs dislocation spacing would be similar to the initial boundary spacing w_0, meaning that the subgrains will tend to be virtually dislocation-free in the beginning of creep. This will increase the mobility of boundaries. As w_0 is smaller than $w_{qs}^{cg}(\sigma)$ in the whole stress interval of creep (Figure 7), the subgrains will tend to coarsen with strain toward the quasi-stationary value w_{qs}^{cg} as long as the grains provide enough space, i.e., as long as w is smaller than

d. The grey area marks the range of possible subgrain coarsening under the simplifying assumptions that all grains are equiaxed and keep the same size $d = d_0$ throughout deformation. The solid black line marks the qs boundary spacing in ufg 8Cu-Zr under these conditions. If this line were exactly valid, all grains would become subgrain-free during deformation for stresses < 270 MPa. However, that is unrealistic because of (i) the wide distribution of HAB spacings (Figure 9) and (ii) dynamic grain growth (see below).

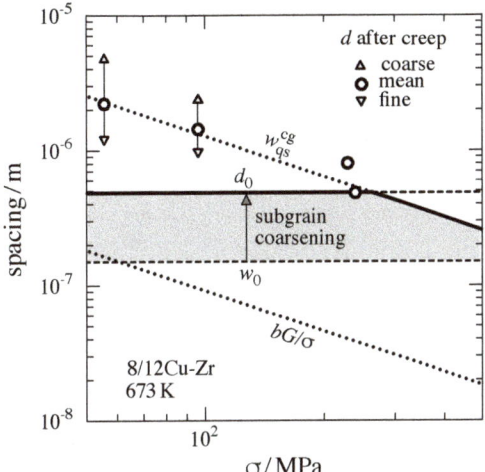

Figure 7. Initial spacings w_0 (all boundaries) and d_0 (HABs only) in pCu-Zr with $p \geq 8$ in comparison to the qs spacings w_{qs}^{cg} of subgrains (Equation (3)) and $\approx bG/\sigma$ of free dislocations expected in coarse-grained Cu-base material; circles: HAB-spacings after deformation, triangles: HAB-spacings in regions with relatively fine and coarse grains, respectively; grey area: range of subgrain coarsening for $d = d_0$.

3.2.2. After Creep

Figure 8 shows grain structures after creep at low σ. There is a mixture of large and fine grain sections. The largest ones are much coarser than in the initial state (Figures 4 and 7). Coarsening is also seen from the TEM micrograph of Figure 5b after fracture at σ_{eng}=150 MPa in comparison to Figure 5a taken before creep. However, this is probably mainly due to subgrains growing from w_0 (Figure 5a) in the attempt to attain the qs spacing w_{qs}^{cg}. Figure 9 shows the distributions of individual HAB-spacings determined from Figures 4 and 8. We will use them as rough experimental input to the model presented in Section 4.

Comparison of Figure 5a,b shows no indications for significant coarsening of particles during creep. So we assume that the stabilization of the grain structure by particles will essentially persist in creep.

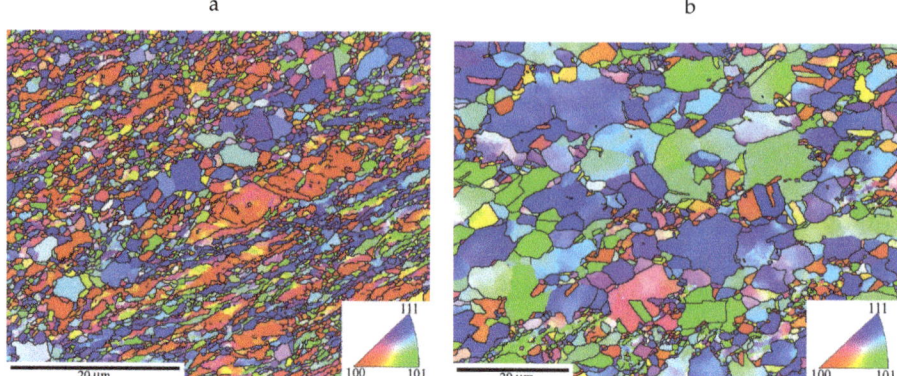

Figure 8. Coarsened grains after slow creep at σ_{eng} = (**a**) 75 MPa and (**b**) 50 MPa (grey triangles in Figure 7); slight color variations in some coarsened grains indicate subgrain formation during creep; insert: color code for crystallographic grain normals in standard triangle.

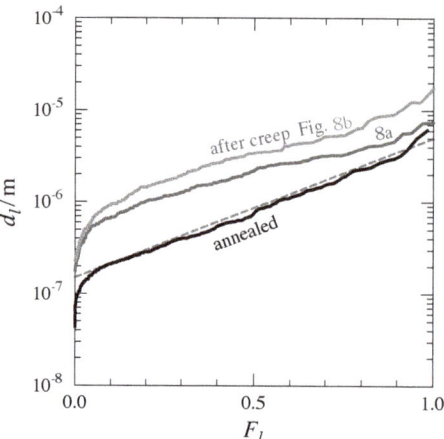

Figure 9. Individual HAB-spacings d_l ($l = 1, 2, \ldots$) versus cumulative spacing fraction F_l evaluated for 8Cu-Zr in annealed state from Figure 4 and after slow creep from Figure 8a,b. Grey dashed line: linear approximation for annealed state.

3.3. Strength Evolution of 8/12Cu-Zr with Strain

Figure 10 displays the $\dot{\epsilon}$-evolution of ufg 8Cu-Zr from the overview of Figure 3e in greater detail as function of strain ϵ. For comparison, the grey dotted lines show the increase of $\dot{\epsilon}$ with ϵ expected for qs deformation with $n_{qs} \approx 6$ at constant σ_{eng}. The scatter in the series of tests at σ_{eng} = 250 MPa indicates that the thermomechanical pretreatment was not perfect and led to some variations of microstructures and strengths. The intermediate σ_{eng}-increase to 83 MPa in the test starting at 75 MPa causes a jump in the qs level of $\dot{\epsilon}$ at $\epsilon \approx 0.15$ (see black squares) which corresponds to n_{qs} = 5.3. This is in reasonable agreement with $n_{qs} \approx 6$. At high (true) stresses σ > 270 MPa and strain rates > $10^{-4}\,\mathrm{s}^{-1}$ the exponent n_{qs} appears to gradually increase with σ beyond 6. This is seen from the test of Figure 11 with a series of load reductions compensating the increase of σ with ϵ (Equation (2)). When the abscissa ϵ in Figure 11a is exchanged for σ in Figure 11b, all the branches with σ_{eng} < 250 MPa shift to the left and superimpose. A common $\dot{\epsilon}$–ϵ line describes both the superimposed curves and their (dashed) connection to the final branch with σ_{eng} = 250 MPa. The slope of the dashed connection is given by n_{qs} = 10.4 at σ > 275 MPa.

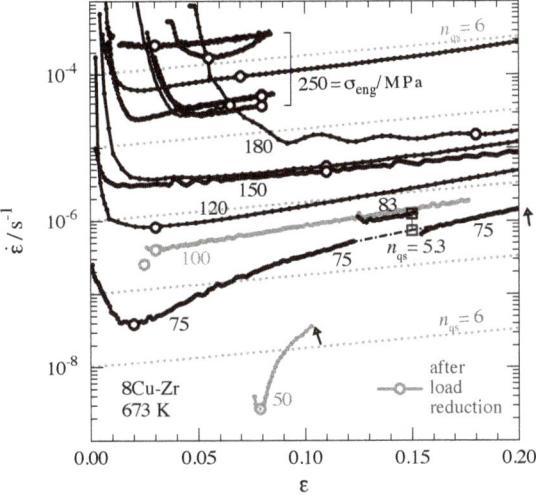

Figure 10. $\dot{\epsilon}$ of 8Cu-Zr as function of ϵ at constant σ_{eng}; black circles: identified qs rates; grey: after σ_{eng}-reduction of from 250 MPa; see Figure 8 for microstructure at black arrows.

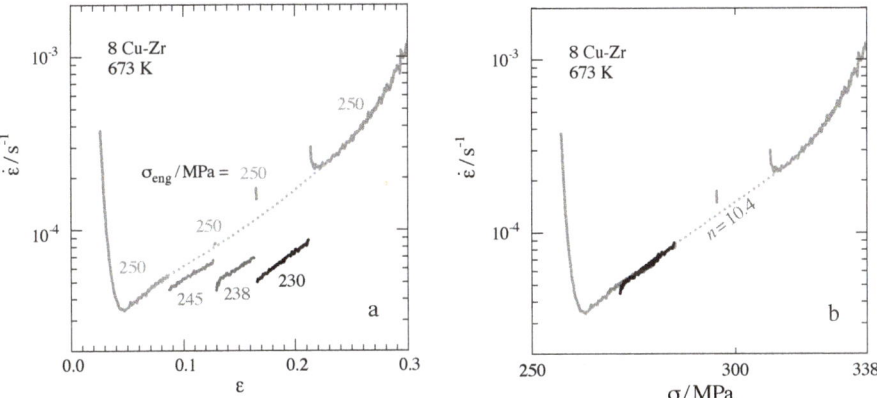

Figure 11. $\dot{\epsilon}$-evolution of 8Cu-Zr as function of (a) ϵ, (b) $\sigma = \sigma_{eng} \exp(\epsilon)$ in test with σ_{eng}-reductions of increasing magnitude to keep $\sigma = \sigma_{eng} \exp(\epsilon)$ about constant.

At low σ<150 MPa the qs strength evolution can no longer be explained only in terms of the slow increase of σ at constant σ_{eng} and fracture (see Figures 3 and 10. Therefore changes in the subgrain/grain structures must be invoked. This corresponds to the observed dynamic coarsening of subgrains and grains (Figure 7, Sections 3.2.1 and 3.2.2).

3.4. Quasi-Stationary Strength of 8/12Cu-Zr

The circles in Figure 10 indicate the quasi-stationary strengths of ufg Cu-Zr at the grain/phase structure present at the beginning of creep. Their choice results from a compromise of

- providing enough strain to fill the material with a qs dislocation structure (spacing of free dislocations and subgrain size) while
- avoiding grain coarsening at large strains.

Taking into account that subgrain growth by migration of LABs is a relatively fast process needing strains in the order of 0.05 to 0.1 and that only limited subgrain growth is possible before all LABs were absorbed at the HABs of mean spacing d_0 (Figure 7), we expect that at the rather low strains of the qs states marked by circles in Figure 10 only marginal grain coarsening will have happened and that sufficiently small grains will have become subgrain-free by subgrain coarsening, except perhaps at the lowest stresses. Figure 12a shows the circles from Figure 10 as function of stress σ for ufg 8Cu-Zr. Within scatter, the qs strengths confirm the power law with a stress exponent $n_{qs} = 6$, already mentioned in Section 3.1, for the stress range 70 MPa $< \sigma <$ 260 MPa. The square symbols from the load change test demonstrate the moderate softening effect resulting from dynamic (sub)grain coarsening mentioned before (Figure 7). An analogous procedure as for 8Cu-Zr leads to Figure 12b for ufg 12Cu-Zr. There is no significant difference between 8Cu-Zr and 12Cu-Zr except lesser scatter for 12Cu-Zr at (true) stresses near 270 MPa (corresponding to $\sigma_{eng} = 250$ MPa), probably due to better homogeneity of the grain structure of 12Cu-Zr.

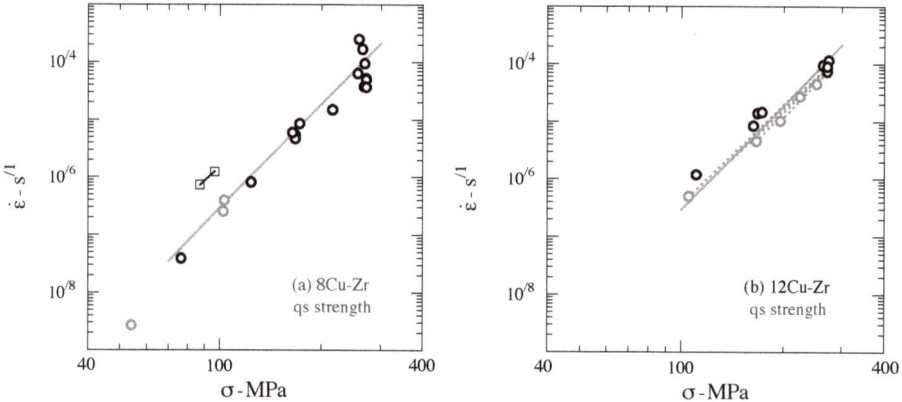

Figure 12. Quasi-stationary strengths of (**a**) 8Cu-Zr (symbols from Figure 7) and (**b**) 12Cu-Zr; grey dotted lines connect data from the same test.

4. Discussion

The primary transients in ufg materials are usually small because the high density of HABs limits the free path of dislocations effectively and leads to high rates of dislocation generation. They are roughly estimated [18] as $d\rho^+/d\varepsilon = 2M/(b\Lambda)$ with $\Lambda \approx d_0 \approx 5 \times 10^{-7}$ m and Taylor factor $M = 3$ for polycrystals. Without concurrent dynamic recovery it needs a strain interval $\Delta\varepsilon^+ \approx M^{-1}(b/d)(\sigma/G)^2 \Delta\rho^+$ to generate the full qs dislocation density $\rho_{qs} \approx b^{-2}(\sigma/G)^2$ in grains of size $d = d_0$. This means that at a stress of 150 MPa a plastic strain of $\Delta\varepsilon^+ = 0.011$ is sufficient to generate the full qs dislocation density. This value is consistent with the primary transients extending over strain intervals > 0.01 in creep at $\sigma_{eng} = 150$ MPa (Figure 10) and confirms that a qs state in the sense of Equation (1) is reached shortly after the $\dot{\varepsilon}$-minimum.

The aim of the present work is to learn more about the influence of the grain structure on the qs strength. As shown above, the high value of f_{HAB} has a softening effect (Section 3.1) and leads to a relatively high rate sensitivity (equivalent to low stress sensitivity n_{qs}) of the qs strength (Section 3.4). The qs stress exponent $n_{qs} \approx 6$ and a high activation energy remind one of the power laws of steady state creep that are explained by specific mechanisms of qs creep, characterized by a certain stress exponent n_{qs} and rate sensitivity $1/n_{qs}$ (see, e.g., [19–21]). For example, mechanisms 1 and 2 in Figure 13a might represent climb-controlled steady state creep and superplastic deformation, respectively. Following this approach, the power law range in Figure 12 would correspond to some mechanism 2.

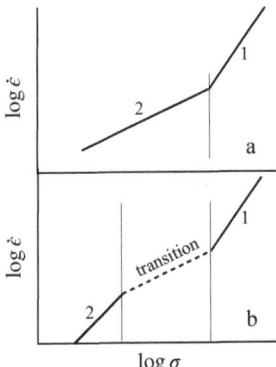

Figure 13. Mechanisms 1 and 2 of qs deformation with (**a**) abrupt, (**b**) smooth transition.

However, there is an alternative possibility [22]. The investigated region with relatively high rate sensitivity may represent a smooth *transition* between a mechanism 1 dominating at high σ and a mechanism 2 dominating at low σ (Figure 13b). A transition region must be expected from the fact that the spacings of HABs have a wide distribution due to coexistence of small and large grains. According to the microstructural data of Figure 7, mechanism 1 would represent qs deformation with subgrain-bearing grains, while mechanism 2 would represent qs deformation with subgrain-free grains (This makes a qualitative difference to Ghosh and Raj [23] who studied the influence of a distribution of grain sizes in relation to the transition between superplastic and normal behavior, but assumed that both mechanisms of deformation are concurrently active in each grain). In the (dashed) transition region both types of grains would be present. In qualitative form this possibility has already been applied to microcrystalline Cu at 0.35 T_m [24]. A semi-quantitative model was provided in [25] and applied to microcrystalline Cu at 0.42 T_m [26]. To apply this model to the present case we need the qs strengths of grains with and without subgrains and the distribution of grain volumes $i = 1, 2, \ldots$ with spacings d_i between the HABs for ufg Cu-Zr at 0.50 T_m. This warrants assumptions based on educated guesses. Following [25,26] we make these choices:

- The cumulative volume fraction F_d with HAB-spacings $\leq d$ is described by the thin straight line in Figure 9 (grey, dashed).
- The qs strength of *subgrain-free* grains is quantified by the relation

$$\dot{\varepsilon} \propto d^4 \sigma^8 \tag{4}$$

from [27]; the f-factors were set to 0.19 (f-factors = 1 apply in the limiting case where all dislocations are lying at HABs, all are in dipolar configuration ready for recovery, and have unrelaxed stress fields; as this is unrealistic, f-values distinctly less than 1 are sensible). This choice yields the two dashed grey lines in Figure 14a for the present F_d and the limiting assumptions of equal stress (iso-stress) or equal strain rate (iso-rate) in all grains.
- The qs strength of crystal volumes *with* subgrains of size $w_{qs}^{cg}(\sigma)$ is estimated by the power law $\dot{\varepsilon} \propto \sigma^{15}$ (dotted line in Figure 14a). The exponent 15 is motivated by the increase of n_{qs} with

stress for $\sigma > 270$ MPa (Figure 12). The position of the line is supported by the result for 2Cu-Zr in Figure 2: The grain size in 2Cu-Zr is so large that all grains contain subgrains; at 150 MPa the upper bound of the initial qs strain rate of 2Cu-Zr is near 10^{-8} s^{-1} (Figures 2 and 3c); this is consistent with the (grey dotted) estimate for subgrain-containing grains in Figure 14a.

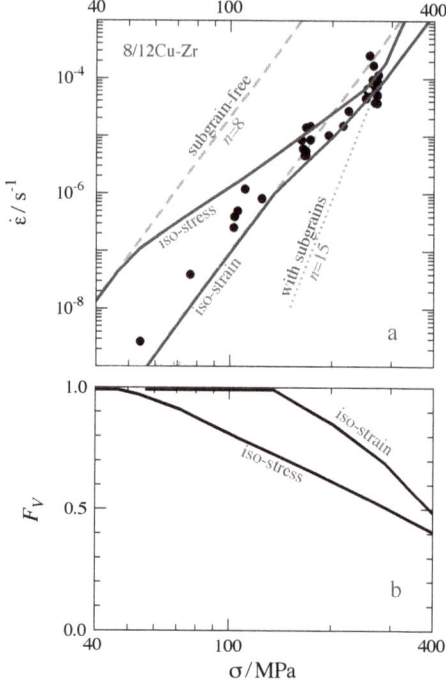

Figure 14. (a) Strain rate and (b) volume fraction of subgrain-free grains as function of stress σ in qs deformation; filled circles: qs data for 8/12Cu-Zr from Figure 12a,b, solid lines: model.

The solid black lines in Figure 14 show the result of the modeling. The iso-rate assumption enforces a redistribution of stress; σ-concentration to hard grains reduces the stresses in soft grains. This stress shielding hinders deformation of large, soft grains and so raises the flow stress level compared to the iso-stress case. The realistic situation lies between the two limits of iso-stress and iso-rate. Comparison of the model lines with the measured data shows reasonable agreement, in particular with regard to the minimal stress sensitivity n_{qs} (maximal rate sensitivity $1/n_{qs}$) in the transition from subgrain-free to subgrain-containing grains.

We note that the difference between the two model lines is rather small in the transition region and that the choice of the solid grey straight F_d-line in Figure 9 has relatively little influence there. The reason may lie in a compensation effect. On the one hand, subgrain-free grains deform faster at a given stress when their grain size d increases (Equation (4)). On the other hand, the fraction of subgrain-free grains decreases with increasing d as more grains develop subgrains. This means that grain coarsening may not always have the dramatic effect expected from the d^4-term in relation (4). At low σ and $\dot\varepsilon$ the situation is unclear because subgrains as well as grains coarsen during creep and the microstructural data are not precise enough for modeling. Therefore, we refrain from a detailed discussion here.

5. Summary

- The quasi-stationary (qs) strength of particle-strengthened ufg Cu-Zr prepared by ECAP and annealing was determined at $0.5\,T_m$ in deformation (creep) at constant load.
- At $\sigma_{eng} = 150\,\text{MPa}$ increase of the fraction f_{HAB} of high-angle boundaries (HABs) leads to softening and ductilization. So HABs have a net softening effect at test conditions.
- The (HAB) grain structure of ufg Cu-Zr is relatively stable at $0.5\,T_m$. At low $\dot{\varepsilon}$ dynamic grain coarsening sets in and leads to softening.
- The initial qs strength of ufg Cu-Zr measured before massive grain coarsening is described by a power law $\dot{\varepsilon} \propto \sigma^n$ with stress exponent $n_{qs} \approx 6$, corresponding to a relatively high rate sensitivity $1/n_{qs} \approx 1/6 = 0.17$, around $\dot{\varepsilon} = 10^{-4}\,\text{s}^{-1}$ and the relatively high stress $\sigma = 170\,\text{MPa} = 4.7 \times 10^{-3}\,G$.
- Analysis of the grain structure indicates that an increasing fraction of small grains becomes subgrain-free in the qs state as the stress decreases.
- The relatively high qs rate sensivity of ufg Cu-Zr is modeled as a result of the variation of the fraction of relatively soft subgrain-free grain volume with stress.

Author Contributions: Conceptualization, W.B. and P.K.; methodology, V.S., J.D., P.K. and P.E.; software, W.B. and P.E.; validation, P.K., J.D.; formal analysis, W.B.; investigation, V.S., J.D., P.K.; resources, J.D.; data curation, J.D., P.K. and W.B.; writing–original draft preparation, W.B.; writing–review and editing, W.B., P.K. and P.E.; visualization, W.B; J.D., P.K., W.B. and P.E.; supervision, W.B. and V.S.

Conflicts of Interest: The authors declare no conflict of interest.

Abbreviations

The following abbreviations are used in this manuscript:

qs	quasi-stationary
ECAP	equal channel angular pressing
ufg	ultrafine-grained
cg	coarse grained
LAB	low-angle boundary
HAB	high-angle boundary
TEM	Transmission electron microscopy

References

1. Blum, W.; Dvořák, J.; Král, P.; Eisenlohr, P.; Sklenička, V. What is stationary deformation of pure Cu? *J. Mater. Sci.* **2014**, *49*, 2987–2997. [CrossRef]
2. Mughrabi, H. Implications of non-negligible microstructural variations during steady state deformation. *Z. Metallkd.* **2005**, *96*, 546–551. [CrossRef]
3. Blum, W.; Eisenlohr, P. Structure evolution and deformation resistance in production and application of nanostructured materials—The concept of steady-state grains. *Mater. Sci. Forum* **2011**, *683*, 163–181. [CrossRef]
4. Valiev, R.Z.; Korznikov, A.V.; Mulyukov, R.R. Structure and properties of ultrafine-grained materials produced by severe plastic deformation. *Mater. Sci. Eng. A* **1993**, *168*, 141–148. [CrossRef]
5. Iwahashi, Y.; Horita, Z.; Nemoto, M.; Langdon, T. The process of grain refinement in equal-channel angular pressing. *Acta Mater.* **1998**, *46*, 3317–3331. [CrossRef]
6. Valiev, R.; Islamgaliev, R.; Alexandrov, I. Bulk nanostructured materials from severe plastic deformation. *Progr. Mater. Sci.* **2000**, *45*, 103–189. [CrossRef]
7. Estrin, Y.; Vinogradov, A. Extreme grain refinement by severe plastic deformation: A wealth of challenging science. *Acta Mater.* **2013**, *61*, 782–817. [CrossRef]
8. Kral, P.; Svoboda, M.; Dvorak, J.; Kvapilova, M.; Sklenicka, V. Microstructure Mechanisms Governing the Creep Life of Ultrafine-grained Cu-0.2wt.%Zr Alloy. *Acta Phys. Pol.* **2012**, *A 122*, 485–489.

9. Dvorak, J.; Kral, P.; Kvapilova, M.; Svoboda, M.; Sklenicka, V. Microstructure stability and creep behaviour of Cu-0.2wt.%Zr alloy processed by equal-channel angular pressing. *Mat. Sci. Forum* **2011**, *667–669*, 821–826. [CrossRef]
10. Sklenička, V.; Dvořák, J.; Král, P.; Svoboda, M.; Kvapilova, M.; Langdon, T. Factors influencing creep flow and ductility in ultrafine – grained metals. *Mater. Sci. Eng. A* **2012**, *558*, 403–411. [CrossRef]
11. Frost, H.J.; Ashby, M.F. *Deformation-Mechanism Maps*; Pergamon Press: Oxford, UK, 1982.
12. Neishi, K.; Horita, Z.; Langdon, T.G. Achieving superplasticity in ultrafine-grained copper: influence of Zn and Zr additions. *Mater. Sci. Eng. A* **2003**, *352*, 129–135. [CrossRef]
13. Li, Y.; Zeng, X.; Blum, W. Transition from strengthening to softening by grain boundaries in ultrafine-grained Cu. *Acta Mater.* **2004**, *52*, 5009–5018. [CrossRef]
14. Amouyal, Y.; Divinski, S.; Estrin, Y.; Rabkin, E. Short-circuit diffusion in an ultrafine-grained copper–zirconium alloy produced by equal channel angular pressing. *Acta Mater.* **2007**, *55*, 5968–5979. [CrossRef]
15. Wongsa-Ngam, J.; Wen, H.; Langdon, T.G. Microstructural evolution in a Cu–Zr alloy processed by a combination of ECAP and HPT. *Mater. Sci. Eng. A* **2013**, *579*, 126–135. [CrossRef]
16. Blum, W. High-Temperature Deformation and Creep of Crystalline Solids. In *Plastic Deformation and Fracture of Materials*; Mughrabi, H., Ed.; VCH Verlagsgesellschaft: Weinheim, Germany, 1993; pp. 359–405.
17. Reppich, B. Particle Strengthening. In *Plastic Deformation and Fracture of Materials*; Mughrabi, H., Ed.; VCH Verlagsgesellschaft: Weinheim, Germany, 1993; pp. 311–357.
18. Li, Y.J.; Mueller, J.; Höppel, H.; Göken, M.; Blum, W. Deformation kinetics of nanocrystalline nickel. *Acta Mater.* **2007**, *55*, 5708–5717. [CrossRef]
19. Mukherjee, A.; Bird, J.; Dorn, J. Experimental Correlations for High-Temperature Creep. *ASM Trans. Q.* **1969**, *62*, 155–179.
20. Čadek, J. *Creep in Metallic Materials*; Elsevier: Amsterdam, The Netherlands, 1988.
21. Kassner, M.; Pérez-Prado, M.T. Five-Power-Law Creep in Single Phase Metals and Alloys. *Prog. Mater. Sci.* **2000**, *45*, 1–102. [CrossRef]
22. Blum, W.; Eisenlohr, P. A simple dislocation model of the influence of high-angle boundaries on the deformation behavior of ultrafine-grained materials. In *15th International Conference on the Strength of Materials (ICSMA-15)*; IOP Publishing: Bristol, UK, 2010; Volume 240, pp. 1–4. [CrossRef]
23. Ghosh, A.K.; Raj, R. Grain size distribution effects in superplasticity. *Acta Metall.* **1981**, *29*, 607–616. [CrossRef]
24. Blum, W.; Dvořák, J.; Král, P.; Eisenlohr, P.; Sklenička, V. Effect of grain refinement by ECAP on creep of pure Cu. *Mater. Sci. Eng. A* **2014**, *590*, 423–432. [CrossRef]
25. Eisenlohr, P.; Blum, W. Maximal strain rate sensitivity of quasi-stationary deformation strength when subgrain size matches grain size. *J. Mater. Sci. Technol.* under review.
26. Blum, W.; Dvořák, J.; Král, P.; Sklenička, V. Dynamic grain coarsening in creep of pure Cu at 0.42 Tm after predeformation by ECAP. *Mater. Sci. Eng. A* **2018**, *731*, 520–529. [CrossRef]
27. Blum, W.; Zeng, X.H. A simple dislocation model of deformation resistance of ultrafine-grained materials explaining Hall–Petch strengthening and enhanced strain rate sensitivity. *Acta Mater.* **2009**, *57*, 1966–1974; Corrigendum to **2011**, *59*, 6205–6206. [CrossRef]

© 2019 by the authors. Licensee MDPI, Basel, Switzerland. This article is an open access article distributed under the terms and conditions of the Creative Commons Attribution (CC BY) license (http://creativecommons.org/licenses/by/4.0/).

Article

Nanoscale Hierarchical Structure of Twins in Nanograins Embedded with Twins and the Strengthening Effect

Haochun Tang [1], Tso-Fu Mark Chang [1,*], Yaw-Wang Chai [2], Chun-Yi Chen [1], Takashi Nagoshi [3], Daisuke Yamane [1], Hiroyuki Ito [1], Katsuyuki Machida [1], Kazuya Masu [1] and Masato Sone [1,*]

- [1] Institute of Innovative Research, Tokyo Institute of Technology, Yokohama 226-8503, Japan
- [2] School of Materials and Chemical Technology, Tokyo Institute of Technology, Yokohama 226-8503, Japan
- [3] National Institute of Advanced Industrial Science and Technology, Tsukuba Ibaraki 305-8564, Japan
- [*] Correspondence: chang.m.aa@m.titech.ac.jp (T.-F.M.C.); msone@pi.titech.ac.jp (M.S.); Tel.: +81-45-924-5631 (T.-F.M.C.)

Received: 6 August 2019; Accepted: 5 September 2019; Published: 6 September 2019

Abstract: Hierarchical structures of 20 nm grains embedded with twins are realized in electrodeposited Au–Cu alloys. The electrodeposition method allows refinement of the average grain size to 20 nm order, and the alloying stabilizes the nanoscale grain structure. Au–Cu alloys are face-centered cubic (FCC) metals with low stacking fault energy that favors formation of growth twins. Due to the hierarchical structure, the Hall–Petch relationship is still observed when the crystalline size (average twin space) is refined to sub 10 nm region. The yield strength reaches 1.50 GPa in an electrodeposited Au–Cu alloy composed of 16.6 ± 1.1 nm grains and the average twin spacing at 4.7 nm.

Keywords: nanotwin; nanograin; Au–Cu alloy; micro-compression; yield strength

1. Introduction

The usage of precious metals in micro-components of microelectromechanical system (MEMS) devices has been demonstrated to allow further enhancement in the sensitivity and miniaturization of the device [1–3]. Among the precious metals, Au is a promising material owing to its advantageous properties and process feasibility in electronic devices [4]. However, concerns regarding the structural stability of gold-based components have been noticed due to the relatively low mechanical strength. Although an improved yield strength (σ_y) of ~500 MPa [5] has been reported by refining the average grain size (d) to nanoscale following the Hall–Petch relationship (HP) [6–8], the strength is still low when compared with materials commonly used in electronic devices. For example, silicon materials are often applied in MEMS devices and possess fracture strength of 1–3 GPa [9]. Besides, enhancement in the strength along with the grain refinement reverses when the average grain size reaches ca. 20 nm [10–13], which is known as the inverse Hall–Petch relationship (iHP). Another strengthening utilizing the HP can be achieved through introduction of twin boundaries into the grains [14], but iHP still occurs when the average twin spacing (λ) reaches ca. 10 nm [15].

In addition to the mechanical properties, there are numerous reports on effects of nanoscale structure on fundamental properties of the material, such as, superconductivity observed in nanostructured $HgBa_2CuO_{4+y}$ [16], La_2CuO_{4+y} [17], and Au–Ag [18]. The phonon density of states of Sn films are reported to be affected by the morphology and grain sizes in nanoscale [19]. Furthermore, electrodeposition is an effective method to control the structures in nanoscale [20].

Enhancement of the mechanical strength by solid solution strengthening can be achieved by alloying of the nanocrystalline Au [11–13]. The yield strength reaches 1.0 GPa in Au–Cu alloys prepared by electrodeposition and evaluated by uniaxial micro-compression tests [21,22]. The high yield strength is a result of synergistic effects of grain boundary and solid solution strengthening mechanisms and the sample size effect [23]. On the other hand, a continuous increase in σ_y of the electrodeposited Au–Cu alloys is observed when the grain size is lower than 10 nm, which is against the iHP reported for Au–Cu alloys when the grain size is in sub 10 nm region [10–13]. The grain sizes reported in previous works were estimated by X-ray diffraction and the Scherrer equation. Grain sizes evaluated by the Scherrer equation are recognized to be close to the real grain sizes observed by transmission electron microscopy (TEM) [24,25] in homogeneous nanocrystalline metals. However, deviations between the Scherrer equation and the TEM results could occur when there is another ordered crystalline structure in the specimen. For instance, twins in face-center cubic (fcc) metals having medium-to-low stacking fault energy (γ_{sf}) are commonly observed, such as gold [26,27] and copper [28], and electrodeposition is an effective method to cause evolution of twins [28–30]. Although there is still no report on formation of twins in Au–Cu alloys, it is necessary to investigate microstructures of the Au–Cu alloys via TEM observation to elucidate the strengthening observed in the iHP region.

Furthermore, the Au–Cu micro-pillar with high copper content shows a gradual decrease in the flow stress just after the yielding point in the stress-strain curve; while the flow stress steadily increases after the yielding for the Au–Cu micro-pillar with a low copper concentration (below 15 at.%) [22]. Such a stress drop phenomenon is rarely reported in nanocrystalline face-centered cubic (fcc) metals and should be clarified.

In this work, formation of twins in the electrodeposited Au–Cu alloys is verified to disclose the continuous strengthening observed in the iHP region. In addition, microstructures of the Au–Cu micro-pillar are evaluated to understand the stress drop observed in the stress–strain curve.

2. Materials and Methods

Au–Cu alloy films were electrodeposited with an electrolyte containing $X_3Au(SO_3)_2$ (X = Na, K) and $CuSO_4$. Details of the electrodeposition procedures are reported in previous studies [21,22]. The chemical composition and crystal structure were characterized by energy-dispersive spectroscopy in a scanning electron microscope (SEM, Hitachi SU4300SE, Tokyo, Japan) and X-ray diffraction (XRD, Rigaku Ultima IV, Tokyo, Japan). For characterization of the mechanical property and in consideration of the sample size effect for MEMS applications, micro-pillars fabricated from the Au–Cu alloy films were prepared. The Au–Cu alloy film electrodeposited specimens were first thinned down to less than 100 μm by mechanical polishing and cut into semicircle disk shapes by a mechanical punch machine. Then micro-pillars with dimensions of $15 \times 15 \times 30$ μm^3 were fabricated using a focus ion beam (FIB, Hitachi FB2100, Tokyo, Japan). Mechanical properties of the Au–Cu alloy micro-pillars were evaluated by micro-compression tests with a displacement-control mode, and the strain rate was 5×10^{-3} s^{-1}. More details of the micro-mechanical testing equipment are described in a previous study [31]. Microstructures of the as-deposited Au–Cu alloys and the deformed micro-pillars were observed using a scanning TEM (STEM, JEOL JEM-2100F, Tokyo, Japan) equipped with a high-resolution TEM (HRTEM) operated at 200 kV. Specimens used in the STEM and TEM were prepared by MultiBeam SEM-FIB (JEOL JIB-4500, Tokyo, Japan). For the deformed specimens, the milling direction of the Ga ion beam in the FIB was parallel to the compression direction.

3. Results and Discussion

Electrodeposited Au–Cu alloys incorporated with nanotwins were confirmed by STEM and HRTEM observation. Figure 1a,b shows the STEM images of the $Au_{85}Cu_{15}$ (15 at.% Cu) and $Au_{68}Cu_{32}$ (32 at.% Cu) alloys, respectively. Individual nanoscale crystal grains and the boundaries could be distinguished from contrasts of the patterns. The average grain sizes were 25.6 ± 4.1 and 16.6 ± 1.1 nm for the $Au_{85}Cu_{15}$ and $Au_{68}Cu_{32}$ alloys, respectively. Nanotwins were observed in the STEM images as indicated by the arrows in Figure 1a,b. Figure 1c shows XRD patterns of the as-electrodeposited $Au_{85}Cu_{15}$ and $Au_{68}Cu_{32}$ alloys. No diffraction peaks from other ordered structure (i.e., $L1_2$ Au_3Cu or $L1_0$ AuCu) were observed except the fcc diffraction peaks, indicating complete solid solution in the electrodeposited Au–Cu alloys. The average sizes of the ordered crystalline estimated by the XRD results and the Scherrer equation were 7.8 and 4.7 nm for the $Au_{85}Cu_{15}$ and $Au_{68}Cu_{32}$ alloys, respectively. The grain sizes observed in STEM (d) were much larger than the average sizes from the Scherrer equation, which implied the average sizes were very likely to be average spacing of the nanotwins (λ). Figure 1d is a representative HRTEM image of the $Au_{85}Cu_{15}$ alloy, which shows a ~30 nm grain containing a ~8 nm wide single band. The electron diffraction patterns converted by fast Fourier transform (FFT) confirmed the nanotwin structure, and the twin is symmetrical to the matrix with the twin boundary (TB) (111) plane. The grain can be divided into three individual bands by the parallel TBs and the widths are all about 10 nm, which is very close to the λ estimated by the Scherrer equation. On the other hand, grains containing only one TB were also observed. As shown in Figure 1e, the TB located in the middle of the grain separates the grain into two equal parts. Illustration of grains divided by one and two TBs is shown in Figure 1f.

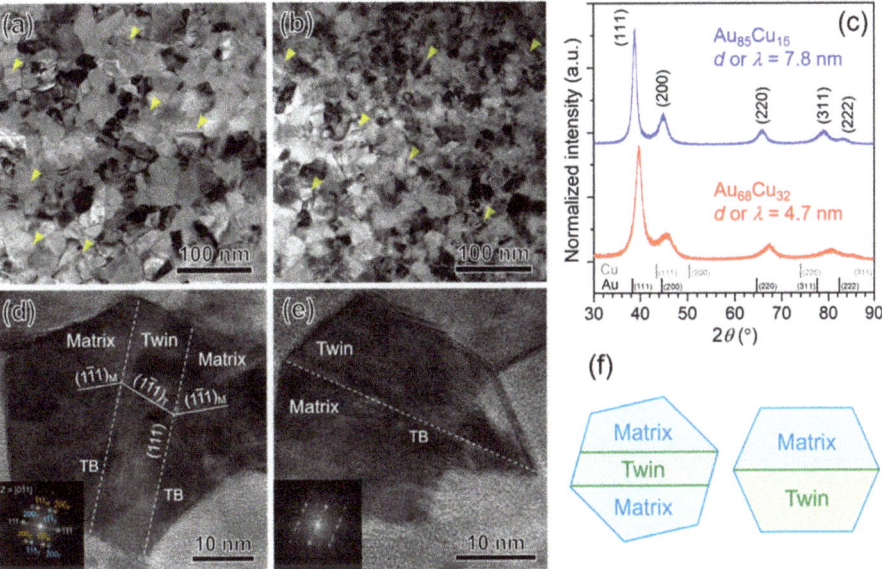

Figure 1. (a,b) Bright-field scanning transmission electron microscopy (STEM) images of as-electrodeposited $Au_{85}Cu_{15}$ and $Au_{68}Cu_{32}$ alloys. The arrows indicate the nanotwins inside the nanograins. (c) XRD patterns of $Au_{85}Cu_{15}$ and $Au_{68}Cu_{32}$ alloys. The vertical bars at bottom indicate the diffraction peaks of pure Au and Cu. (d,e) Two representative high-resolution transmission electron microscopy (HRTEM) images taken from the $Au_{85}Cu_{15}$ alloy. Zone axis: [0$\bar{1}$1]. The vertical bars at bottom indicate the diffraction peaks of pure Au and Cu. (f) Illustration of two types of the nanotwin in a nanograin.

For alloy electrodeposition, the applied current density plays an important role in controlling the grain size and composition. In the case of Au–Cu alloys, the Cu concentration is increased by applying a higher cathodic current density due to the difference in standard reduction potential between Au and Cu ions [11,13]. Meanwhile, the higher current density can promote the nucleation rate resulting in finer grains in electrodeposits [32]. The twin evolution is attributed to the lowered γ_{sf} by alloying two fcc metals already with relatively low γ_{sf}. A strong decrease in the γ_{sf} as a result of alloying was experimentally examined and revealed to have a semi-log relationship in most of fcc-based alloys (i.e., Ag, Cu, Ni) as expressed in the following [33,34]:

$$\ln\frac{\gamma_{sf}}{\gamma_0} = k_\gamma \left(\frac{x}{1+x}\right)^2, \quad (1)$$

where γ_0 is the stacking fault energy of the solvent metal. k_γ is a material constant. x is the expression of c/c^*, where c is the solute concentration, and c^* is the solubility limit. For example, the stacking fault energy of pure Cu reduces from ~70 mJ/m^2 to a value lower than 10 mJ/m^2 when forming Cu-based alloys [33]. Wu et al. reported the formation of a nanotwinned structure in electrodeposited Ni–80Co alloys with average grain size of ~30 nm, and the growth twins were reported to be affected by γ_{sf} of the alloy [29]. Lucadamo et al. also observed the twinning features in electrodeposited Ni–Mn alloys but with coarser grains of ~200 nm [30].

Micro-mechanical properties of the electrodeposited Au–Cu alloys were revealed by micro-compression tests. Figure 2a–d shows SEM images of the as-fabricated Au$_{85}$Cu$_{15}$ and Au$_{68}$Cu$_{32}$ micro-pillars and after compression with 12%–14% plastic strain. Similar barrel-shape deformations were observed in both micro-pillars, which were typical deformation behaviors for polycrystalline metallic materials. The engineering stress–strain curves obtained from the compression tests are shown in Figure 2e. The σ_y's of the Au$_{85}$Cu$_{15}$ and Au$_{68}$Cu$_{32}$ micro-pillars were 0.95 and 1.16 GPa, respectively. After the yielding point, the Au$_{85}$Cu$_{15}$ pillar exhibited a steady increase in the flow stress during the plastic deformation until unloading. For the Au$_{68}$Cu$_{32}$ pillar, the flow stress declined in the early stage of the plastic deformation for strain of ~2%. After that, the flow stress steadily increased similar to that of the Au$_{85}$Cu$_{15}$ pillar. It should be noticed that there is still no report on the stress drop for pure polycrystalline fcc micro-specimens.

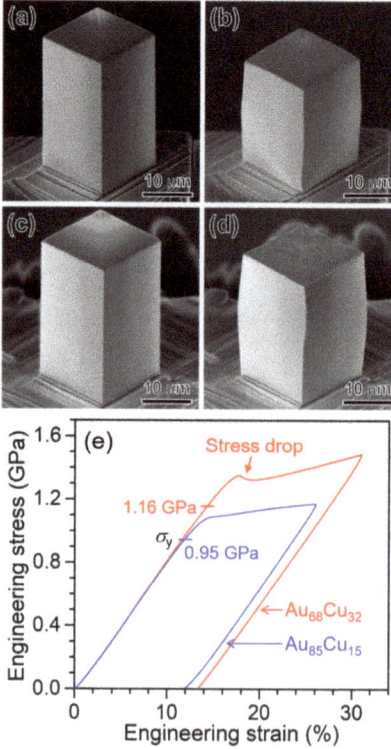

Figure 2. SEM images of (**a,b**) Au$_{85}$Cu$_{15}$ and (**c,d**) Au$_{68}$Cu$_{32}$ micro-pillars (**a,c**) before and (**b,d**) after the compression with 12%–14% strain. (**e**) Engineering stress–strain curves obtained from the micro-compression tests.

To understand the stress drop observed in the stress–strain curves, microstructures of the deformed micro-pillars were further investigated by the STEM and HRTEM. Figure 3a shows a STEM image of the $Au_{68}Cu_{32}$ alloy after compression of 13.8% plastic strain. Similar to the as-electrodeposited alloys shown in Figure 1a,b, conspicuous nanotwins were observed inside the nanograins. In addition to the growth twins, deformation twins inside highly deformed grains were observed as shown in Figure 3b. In the image, a deformation TB next to a growth TB located at the left side of the grain was observed, and the deformation TB was obstructed in the middle of the grain. Another deformation TB could be observed at the right side of the grain. A magnified inverse fast Fourier transform (IFFT) image shown in Figure 3c reveals the extremely complex interaction between the deformation twin and dislocation, which forces the twinning to be interrupted inside the grain.

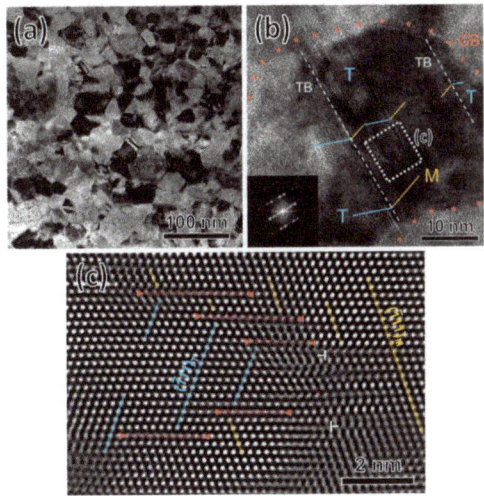

Figure 3. (a) A STEM image of the $Au_{68}Cu_{32}$ micro-pillar after ~13.8% compressive strain. (b) An HRTEM image of a highly deformed grain showing deformation twin. T: twin, M: matrix, Zone axis: [0$\bar{1}$1], and (c) a magnified inverse fast Fourier transform (IFFT) image showing the deformation twin.

Deformation twinning is one of major deformation mechanisms not only in fcc metals with low γ_{sf}, but also in nanocrystalline fcc metals with high γ_{sf} if deformed under extreme conditions [35,36]. Several deformation twinning mechanisms are proposed and observed in nanocrystalline fcc metals, i.e., the random activation of partials mechanism [37], the dislocation rebound mechanism [38], or the partial emissions from grain boundary [35,38]. When a twin structure initiates from the grain boundary and terminates inside a grain, it can only be formed by the partial emissions from grain boundary. Zhu et al. [39] observed similar results in nanocrystalline Ni and proposed the relative mechanisms for Shockley twinning partials to multiply at grain boundary (GB). Furthermore, the γ_{sf} of the fcc metals is usually reduced by alloying, especially for the Au–Cu alloys. Therefore, the reduction in γ_{sf} can change the energy path (i.e., general planar fault energy [40]) and, thus, facilitates the deformation twinning under the applied stress. The stress drop observed in the $Au_{68}Cu_{32}$ alloy pillar is reasonably considered to be the lowered energy requirement for Shockley twinning partials threading into grains to form deformation twins.

Mechanical strengths of polycrystalline metals are often affected by multiple strengthening mechanisms taking place simultaneously. In the present case of electrodeposited Au–Cu alloys, the obtained σ_y are considered to be the synergistic effects of grain boundary strengthening, twin boundary strengthening, and solid solution strengthening. The effect of grain size on the strength is known to be the Hall–Petch relationship [6,7]:

$$\sigma_{gb} = \sigma_0 + k_{HP} \cdot d^{-1/2}, \tag{2}$$

where σ_{gb} is the strength contributed from GB, σ_0 is the friction resistance for dislocation movement within the polycrystalline grains, k_{HP} is the Hall–Petch coefficient, and d is the grain size. The twin boundary could form barriers to the dislocation motion similar to the grain boundary. Lu et al. [15,28,41] reported that the average twin width (λ) and strength of the specimen follows a Hall–Petch relationship-like behavior in the columnar-grained Cu with high density nanotwins perpendicular to the growth direction. On the other hand, the nanotwin in columnar grain structure is different from the ones present Au–Cu alloys. The Au–Cu alloys evaluated in this study were composed of isotropic grains of much smaller grain size, and because of the ~20 nm average grain size, each grain could accommodate a low number (mostly one and two in this study) of the twin boundaries and resulted a sub 10 nm average twin width. Figure 4a shows the Hall–Petch plot for Au–Cu alloys including the results of the present study and literatures evaluated by Vickers hardness tests [10,12,13]. σ_y of the Au–Cu alloys increased from 0.90 to 1.50 GPa, when the λ decreased from 4.7 to 9.1 nm.

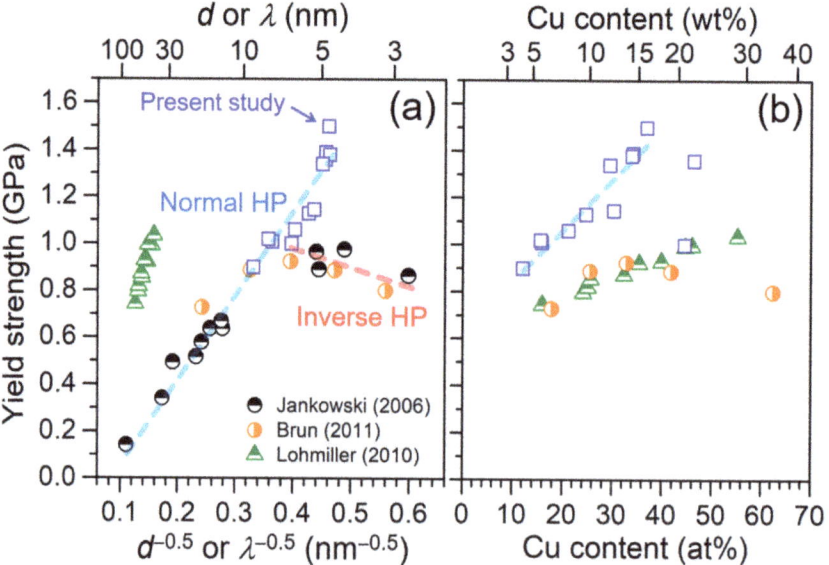

Figure 4. (a) Hall–Petch plot of Au–Cu alloys. The yield strengths in the literatures were converted from Vickers hardness. (b) A plot of yield strengths as a function of Cu content.

For the solid solution strengthening, the classical theories are well established in coarse-grained alloys such as Fleischer model [42] and Labusch theory [43]. Rupert and Schuh et al. further proposed enhanced models for nanocrystalline fcc alloys, in which the σ_y and the strength contributed from nanocrystalline solid solution ($\Delta\sigma_{nc,SS}$) are expressed by [44,45]:

$$\sigma_y = A \cdot E, \tag{3}$$

$$\Delta\sigma_{ns,SS} = A \cdot \left(\frac{\partial E}{\partial c}\right) \cdot C, \tag{4}$$

where A is a fitting constant having a function of the applied strain rate and grain size, E is elastic modulus of the alloy, c is composition in at.%. Equations (3) and (4) suggested that the strength in nanocrystalline alloys is not only dominated by the grain size but also affected by the elastic modulus and composition. The copper concentration of the Au–Cu alloys prepared in this study ranged from 12.1 to 46.4 at.%. Here, we assume two conditions to approach the constant A: (i) grain sizes in all Au–Cu alloys are similar and (ii) E follows a linear fashion with alloy composition and ranges between the elastic modulus of Au (74 GPa) and Cu (117 GPa). By doing the assumptions, the fitting constant A is equal to 0.0375, which is somewhat larger than the value reported for nanocrystalline Cu alloys (0.024) [45]. Nevertheless, this modified model for nanocrystalline alloys is in line with our experimental results as shown in Figure 4b.

Au–Cu alloys prepared in this study were confirmed to have ~20 nm as the average grain size and sub 10 nm as the average twin spacing. Both values were still in the HP region and close to the critical value for occurrence of the iHP, which demonstrated thorough utilization of the HP in strengthening of Au–Cu alloys. Due to this, an ultrahigh yield strength of 1.5 GPa was obtained.

4. Conclusions

A hierarchical nanostructure of nanocrystalline Au–Cu alloys containing nanotwins was produced by electrodeposition from sulfite-based electrolyte. Microstructure investigation revealed that average grain sizes of the alloys were about 20 nm, and twin boundaries were observed in the nanograins. Due to the fine grain size, average spacings of the twins were all less than 10 nm, and this confirmed continuous strengthening was observed when the average twin spacing is thinned downed to sub 10 nm region. By making a hierarchical structure of twinned nanograins having the size in the HP region but close to the iHP region, a high yield strength of 1.5 GPa was obtained. In addition, the stress drop observed in the stress–strain curve was caused by evolution of the deformation twins, and the deformation twins were formed because of the reduced stacking fault energy in the Au–Cu alloys.

Author Contributions: Conceptualization, H.T., T.-F.M.C., and M.S.; data curation, H.T. and T.-F.M.C.; validation, H.T., C.-Y.C., H.I., and D.Y.; formal analysis, H.T. and T.N.; investigation, H.T. and Y.-W.C.; resources, M.S.; writing—original draft preparation, H.T.; writing—review and editing, T.-F.M.C.; visualization, H.T.; supervision, T.-F.M.C. and M.S.; project administration, K.M. (Katsuyuki Machida) and M.S.; funding acquisition, K.M. (Kazuya Masu) and M.S.

Funding: This work was supported by JST CREST grant number JPMJCR1433, Japan and the Grant-in-Aid for Scientific Research (S) (JSPS KAKENHI grant number 26220907).

Acknowledgments: The authors thank Suzukakedai Materials Analysis Division, Technical Department, Tokyo Institute of Technology, for TEM specimen preparation.

Conflicts of Interest: The authors declare no conflict of interest.

References

1. Mayagoitia, R.E.; Nene, A.V.; Veltink, P.H. Accelerometer and rate gyroscope measurement of kinematics: An inexpensive alternative to optical motion analysis systems. *J. Biomech.* **2005**, *35*, 537–542. [CrossRef]
2. Yamane, D.; Konishi, T.; Matsushima, T.; Machida, K.; Toshiyoshi, H.; Masu, K. Design of sub-1g microelectromechanical systems accelerometers. *Appl. Phys. Lett.* **2014**, *104*, 074102. [CrossRef]
3. Coutu, R.A.; Kladitis, P.E.; Cortez, R.; Strawser, R.E.; Crane, R.L. Micro-switches with sputtered Au, AuPd, Au-on-AuPt, and AuPtCu alloy electric contacts. *IEEE Trans. Compon. Packag. Technol.* **2016**, *29*, 341–349. [CrossRef]
4. Green, T.A. Gold electrodeposition for microelectronic, optoelectronic and microsystem applications. *Gold Bull.* **2007**, *40*, 105–114. [CrossRef]
5. Tang, H.; Chen, C.Y.; Nagoshi, T.; Chang, T.F.M.; Yamane, D.; Machida, K.; Masu, K.; Sone, M. Enhancement of mechanical strength in Au films electroplated with supercritical carbon dioxide. *Electrochem. Commun.* **2016**, *72*, 126–130. [CrossRef]
6. Hall, E.O. The deformation and ageing of mild steel: III Discussion of results. *Proc. Phys. Soc. London, Sect. B* **1951**, *64*, 747–753. [CrossRef]
7. Petch, N.J. The cleavage strength of polycrystals. *J. Iron Steel Inst.* **1953**, *174*, 25–28.

8. Armstrong, R.W. Hall-Petch description of nanopolycrystalline Cu, Ni and Al strength levels and strain rate sensitivities. *Philos. Mag.* **2016**, *96*, 3097–3108. [CrossRef]
9. Tsuchiya, T.; Tabata, O.; Sakata, J.; Taga, Y. Specimen size effect on tensile strength of surface-micromachined polycrystalline silicon thin films. *J. Microelectromech. Syst.* **1998**, *7*, 106–113. [CrossRef]
10. Schiøtz, J.; Vegge, T.; Di Tolla, F.D.; Jacobsen, K.W. Atomic-scale simulations of the mechanical deformation of nanocrystalline metals. *Phys. Rev. B* **1999**, *60*, 11971–11983. [CrossRef]
11. Jankowski, A.F.; Saw, C.K.; Harper, J.F.; Vallier, B.F.; Ferreira, J.L.; Hayes, J.P. Nanocrystalline growth and grain-size effects in Au–Cu electrodeposits. *Thin Solid Films* **2006**, *494*, 268–273. [CrossRef]
12. Lohmiller, J.; Woo, N.C.; Spolenak, R. Microstructure–property relationship in highly ductile Au–Cu thin films for flexible electronics. *Mater. Sci. Eng. A* **2010**, *527*, 7731–7740. [CrossRef]
13. Brun, E.; Durut, F.; Botrei, R.; Theobald, M.; Legaie, O.; Popa, I.; Vignal, V. Influence of the electrochemical parameters on the properties of electroplated Au-Cu alloys. *J. Electrochem. Soc* **2011**, *158*, D223–D227. [CrossRef]
14. Lu, K. Stabilizing nanostructures in metals using grain and twin boundary architectures. *Nat. Rev. Mater.* **2016**, *1*, 1–13. [CrossRef]
15. Lu, L.; Chen, X.; Huang, X.; Lu, K. Revealing the maximum strength in nanotwinned copper. *Science* **2009**, *323*, 607–610. [CrossRef]
16. Campi, G.; Bianconi, A.; Poccia, N.; Bianconi, G.; Barba, L.; Arrighetti, G.; Innocenti, D.; Karpinski, J.; Zhigadlo, N.D.; Kazakov, S.M.; et al. Inhomogeneity of charge-density-wave order and quenched disorder in a high-T_c superconductor. *Nature* **2015**, *525*, 359–362. [CrossRef]
17. Campi, G.; Bianconi, A. Evolution of complexity in out-of-equilibrium systems by time-resolved or space-resolved synchrotron radiation techniques. *Condens. Matter.* **2019**, *4*, 32. [CrossRef]
18. Thapa, D.K.; Islam, S.; Saha, S.K.; Mahapatra, P.S.; Bhattacharyya, B.; Sai, T.P.; Mahadevu, R.; Patil, S.; Ghosh, A.; Pandey, A. Coexistence of diamagnetism and vanishingly small electrical resistance at ambient temperature and pressure in nanostructures. *arXiv Prepr.* **2018**, *1807*, 08572.
19. Houben, K.; Couet, S.; Trekels, M.; Menéndez, E.; Peissker, T.; Seo, J.W.; Hu, M.Y.; Zhao, J.Y.; Alp, E.E.; Roelants, S.; et al. Lattice dynamics in Sn nanoislands and cluster-assembled films. *Phys. Rev. B.* **2017**, *95*, 155413. [CrossRef]
20. Xiao, Z.L.; Han, C.Y.; Kwok, W.K.; Wang, H.H.; Welp, U.; Wang, J.; Crabtree, G.W. Tuning the architecture of mesostructures by electrodeposition. *J. Am. Chem. Soc.* **2004**, *126*, 2316–2317. [CrossRef]
21. Tang, H.; Chen, C.Y.; Yoshiba, M.; Nagoshi, T.; Chang, T.F.M.; Yamane, D.; Machida, K.; Masu, K.; Sone, M. High-strength electroplated Au-Cu alloys as micro-components in MEMS devices. *J. Electrochem. Soc* **2017**, *164*, D244–D247. [CrossRef]
22. Tang, H.; Chen, C.Y.; Nagoshi, T.; Chang, T.F.M.; Yamane, D.; Konishi, T.; Machida, K.; Masu, K.; Sone, M. Au-Cu Alloys prepared by pulse electrodeposition toward applications as movable micro-components in electronic devices. *J. Electrochem. Soc.* **2018**, *165*, D58–D63. [CrossRef]
23. Uchic, M.D.; Dimiduk, D.M.; Florando, J.N.; Nix, W.D. Sample dimensions influence strength and crystal plasticity. *Science* **2004**, *305*, 986–989. [CrossRef]
24. Zhang, Z.; Zhou, F.; Lavernia, E.J. On the analysis of grain size in bulk nanocrystalline materials via x-ray diffraction. *Metall. Mater. Trans. A* **2003**, *34*, 1349–1355. [CrossRef]
25. Nagoshi, T.; Chang, T.F.M.; Sato, T.; Sone, M. Mechanical properties of nickel fabricated by electroplating with supercritical CO_2 emulsion evaluated by micro-compression test using non-tapered micro-sized pillar. *Microelectron. Eng.* **2013**, *110*, 270–273. [CrossRef]
26. Krajčí, M.; Kameoka, S.; Tsai, A.P. Twinning in fcc lattice creates low-coordinated catalytically active sites in porous gold. *J. Chem. Phys.* **2016**, *145*, 084703. [CrossRef]
27. Lee, S.; Im, J.; Yoo, Y.; Bitzek, E.; Kiener, D.; Richter, G.; Kim, B.; Oh, S.H. Reversible cyclic deformation mechanism of gold nanowires by twinning–detwinning transition evidenced from in situ TEM. *Nat. Commun.* **2014**, *5*, 3033. [CrossRef]
28. Lu, L.; Shen, Y.; Chen, X.; Qian, L.; Lu, K. Ultrahigh strength and high electrical conductivity in copper. *Science* **2004**, *304*, 422–426. [CrossRef]
29. Wu, B.Y.C.; Schuh, C.A.; Ferreira, P.J. Nanostructured Ni-Co alloys with tailorable grain size and twin density. *Metall. Mater. Trans. A* **2005**, *36*, 1927–1936. [CrossRef]

30. Lucadamo, G.; Medlin, D.L.; Yang, N.Y.C.; Kelly, J.J.; Talin, A.A. Characterization of twinning in electrodeposited Ni–Mn alloys. *Philos. Mag.* **2005**, *85*, 2549–2560. [CrossRef]
31. Takashima, K.; Higo, Y.; Sugiura, S.; Shimojo, M. Fatigue crack growth behavior of micro-sized specimens prepared from an electroless plated Ni-P amorphous alloy thin film. *Mater. Trans.* **2001**, *42*, 68–73. [CrossRef]
32. Natter, H.; Krajewski, T.; Hempelmann, R. Nanocrystalline palladium by pulsed electrodeposition. *Ber. Bunsenges. Phys. Chem.* **1996**, *100*, 55–64. [CrossRef]
33. Gallagher, P.C.J. The influence of alloying, temperature, and related effects on the stacking fault energy. *Metall. Trans.* **1971**, *1*, 2429–2461.
34. Hong, S.I.; Laird, C. Mechanisms of slip mode modification in F.C.C. solid solutions. *Acta Metall. Mater.* **1990**, *38*, 1581–1594. [CrossRef]
35. Yamakov, V.; Wolf, D.; Phillpot, S.R.; Mukherjee, A.K.; Gleiter, H. Dislocation processes in the deformation of nanocrystalline aluminium by molecular-dynamics simulation. *Nat. Mater.* **2002**, *1*, 45–49. [CrossRef]
36. Zhu, Y.T.; Liao, X.Z.; Wu, X.L. Deformation twinning in nanocrystalline materials. *Prog. Mater Sci.* **2012**, *57*, 1–62. [CrossRef]
37. Wu, X.L.; Liao, X.Z.; Srinivasan, S.G.; Zhou, F.; Lavernia, E.J.; Valiev, R.Z.; Zhu, Y.T. New deformation twinning mechanism generates zero macroscopic strain in nanocrystalline metals. *Phys. Rev. Lett.* **2008**, *100*, 095701. [CrossRef]
38. Zhu, Y.T.; Narayan, J.; Hirth, J.P.; Mahajan, S.; Wu, X.L.; Liao, X.Z. Formation of single and multiple deformation twins in nanocrystalline fcc metals. *Acta Mater.* **2009**, *57*, 3763–3770. [CrossRef]
39. Zhu, Y.T.; Wu, X.L.; Liao, X.Z.; Narayan, J.; Mathaudhu, S.N.; Kecskés, L.J. Twinning partial multiplication at grain boundary in nanocrystalline fcc metals. *Appl. Phys. Lett.* **2009**, *95*, 031909. [CrossRef]
40. Kibey, S.; Liu, J.B.; Johnson, D.D.; Sehitoglu, H. Predicting twinning stress in fcc metals: Linking twin-energy pathways to twin nucleation. *Acta Mater.* **2007**, *55*, 6843–6851. [CrossRef]
41. Shen, Y.F.; Lu, L.; Lu, Q.H.; Jin, Z.H.; Lu, K. Tensile properties of copper with nano-scale twins. *Scr. Mater.* **2005**, *52*, 989–994. [CrossRef]
42. Fleischer, R.L. Substitutional solution hardening. *Acta Metall.* **1963**, *11*, 203–209. [CrossRef]
43. Labusch, R. A Statistical theory of solid solution hardening. *Phys. Status Solidi B* **1970**, *41*, 659–669. [CrossRef]
44. Rupert, T.J.; Trenkle, J.C.; Schuh, C.A. Enhanced solid solution effects on the strength of nanocrystalline alloys. *Acta Mater.* **2011**, *59*, 1619–1631. [CrossRef]
45. Rupert, T.J. Solid solution strengthening and softening due to collective nanocrystalline deformation physics. *Scr. Mater.* **2014**, *81*, 44–47. [CrossRef]

© 2019 by the authors. Licensee MDPI, Basel, Switzerland. This article is an open access article distributed under the terms and conditions of the Creative Commons Attribution (CC BY) license (http://creativecommons.org/licenses/by/4.0/).

Article

From Statistical Correlations to Stochasticity and Size Effects in Sub-Micron Crystal Plasticity

Hengxu Song [1] and Stefanos Papanikolaou [1,2,*]

[1] Department of Mechanical and Aerospace Engineering, West Virginia University, Morgantown, WV 26506, USA
[2] Department of Physics & Astronomy, West Virginia University, Morgantown, WV 26506, USA
* Correspondence: stefanos.papanikolaou@mail.wvu.edu

Received: 24 June 2019; Accepted: 25 July 2019; Published: 27 July 2019

Abstract: Metals in small volumes display a strong dependence on initial conditions, which translates into size effects and stochastic mechanical responses. In the context of crystal plasticity, this amounts to the role of pre-existing dislocation configurations that may emerge due to prior processing. Here, we study a minimal but realistic model of uniaxial compression of sub-micron finite volumes. We show how the statistical correlations of pre-existing dislocation configurations may influence the mechanical response in multi-slip crystal plasticity, in connection to the finite volume size and the initial dislocation density. In addition, spatial dislocation correlations display evidence that plasticity is strongly influenced by the formation of walls composed of bound dislocation dipoles.

Keywords: plasticity; representative volume element; dislocation structure; dislocation correlations; dislocation avalanches

1. Introduction

Crystal plasticity modelling of a macroscopic cylinder typically requires elasto-plastic constitutive laws. Usually, the onset of crystal plasticity is modeled through a smooth, continuous transformation [1,2], even though in the *rare* absence of pre-existing *mobile* defects it is a fact that the plasticity transition is discontinuous (see Figure 1). In contrast, during nanopillar compression, mobile defects are suggested to be absent [3,4] and the transition is characterized by discontinuous abrupt event sequences (nanoscale) [3,5–15]. Naively, one might expect that the averaging of an abrupt nanopillar response would lead to a discontinuous average response at the nominal yield point. However, unconventional size-dependent nonlinear ensemble average behavior emerges during quasi-static nanopillar compression of crystals as size decreases [16,17].

In uniaxial compression of microscopic crystals, discontinuous plastic yielding may be realized by considering a collection of randomly placed dislocation sources (pinned dislocation segments) in an otherwise dislocation-free crystal (see Figure 1). However, even in such an idealistic case, after loading to a finite strain, the unloading process to zero stress will leave a corresponding plastic strain and dislocation structure. Reloading to the flow stress appears quasi-continuous, but the behavior is typically nonlinear and "anelastic" [18–24], originating in locally irreversible but small deformations that correspond to abrupt jumps of pre-existing dislocations. Experimentally in small volumes, it has been found that uniaxial compression of crystalline nanopillars ranging from ∼100 nm to ∼10 µm is characterized by the absence of mobile dislocation segments ("exhaustion" mechanisms), leading to abrupt events and jerky loading responses [7,25–28].

The ensemble average of small-volume abrupt behavior, smooth and nonlinear, resembles macroscale crystal plasticity. For uniaxial compression of cylinders, due to the absence of geometric gradients, it is natural to consider crystal plasticity a a *local* phenomenon [29]. Thus, it is expected that the ensemble average of nanopillar responses should equal the *spatial average* response of a macroscopic cylinder. Nevertheless, recent experiments [16,17] displayed strong size dependence for the average mechanical response of copper single crystalline pillars with sizes decreasing from 3 µm to 300 nm, showing increasing curvature during quasi-static loading [1,30–34]. At which scale does the micropillar statistical ensemble averaged strength and hardening equal the spatially self-averaged ones?

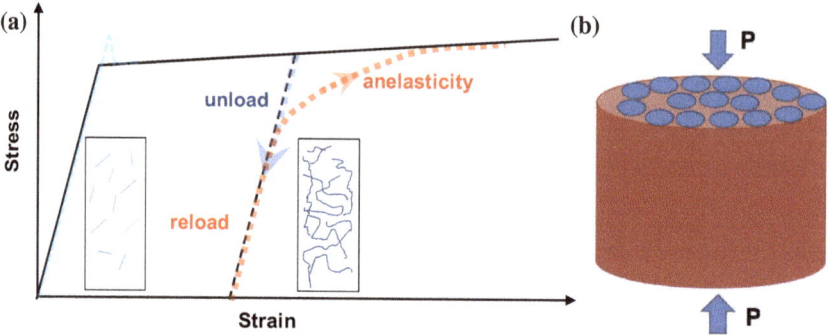

Figure 1. Schematic of ensemble and spatial averaging in uniaxial compression of cylindrical samples. (**a**) An annealed small volume yet with dislocation sources (pinned dislocation segments), loaded to a certain strain, responds abruptly at the elastic-plastic transition (solid:load-control, dashed:displacement-control). If the sample is pre-deformed, reloading (dashed line, reload) shows nonlinearity before reaching the flow stress. (**b**) The uniaxial compression of a cylinder, at load **P**, may be thought as a statistical collection of microscale pillars' compression. However, at what scale should each pillar compression be considered for such an averaging to be accurate?

In this paper, we investigate how the statistical ensemble average of plastic, abrupt mechanical response of uniaxially stressed small volumes depends on the system size and pre-existing dislocation microstructure. We perform an explicit but minimal discrete dislocation dynamics model study with one and two active slip systems. Two typical initial dislocation microstructures are utilized: (i) annealed (dislocation free) samples; (ii) "mobile-dislocation-rich" dislocation microstructures created by a prior loading history. We demonstrate that the onset of plasticity and continuous nonlinearity of stress–strain curves is caused by inhomogeneous dislocation microstructures that form under prior multislip loading, composed of dislocation dipoles. We also show that, in this model of uniaxial compression, the very observation of *scale free* power law avalanche behavior is connected to the emergence of the statistically averaged stress–strain curvature. Based on this model evidence, we conclude that single-slip plasticity may be ensemble averaged by compressed nanopillars with diameters even less than 500 nm. However, multi-slip plasticity may be averaged only by finite volume pillar compression with volumes larger than 2–4 µm.

The paper is organized as follows: Section 2 contains the model description and details of our study; Section 3 is focused on the mechanical response of nanopillars of different sizes and microstructures for multi-slip conditions. In Section 4, we focus on the nonlinearity of statistical ensemble average and its connection to spatial edge dislocation-pair correlations. Avalanche statistics is also discussed for different dislocation densities. In Section 5, we discuss our conclusions in the context of the macroscopic constitutive relations derived by the small-volume response ensembles.

2. Model Description

The uniaxial compression of a nano/micro-pillar is carried out by two-dimensional (2D) discrete dislocation dynamics, where only edge dislocations are considered in one or multiple slip systems. This is an accurate model for thin films [11,12] and it can be considered as a phenomenologically consistent model for uniaxial nanopillar compression [7,35]. The schematic of the uniaxial compression is shown in Figure 2. Using small strain assumptions, plastic deformation is described through the framework developed in [36], where the material's state determination employs strain/stress superposition. Thus, shape asymmetries related to plastic deformation are effectively not considered. Each edge dislocation is treated as a singularity in an infinite space with Young modulus E and Poisson ratio ν. The application of the dislocation analytical solution, which is valid in an infinite space, needs a smooth image field (^) to ensure that actual boundary conditions are satisfied. Hence, the displacements u_i, strains ε_{ij}, and stresses σ_{ij} are written as

$$u_i = \tilde{u}_i + \hat{u}_i, \quad \varepsilon_{ij} = \tilde{\varepsilon}_{ij} + \hat{\varepsilon}_{ij}, \quad \sigma_{ij} = \tilde{\sigma}_{ij} + \hat{\sigma}_{ij}, \quad (1)$$

where the (̃) field is the sum of the fields of all N dislocations in their current positions, i.e.,

$$\tilde{u}_i = \sum_{J=1}^{N} \tilde{u}_i^{(J)}, \quad \tilde{\varepsilon}_{ij} = \sum_{J=1}^{N} \tilde{\varepsilon}_{ij}^{(J)}, \quad \tilde{\sigma}_{ij} = \sum_{J=1}^{N} \tilde{\sigma}_{ij}^{(J)}. \quad (2)$$

The image fields ^ are obtained by solving a linear elastic boundary value problem using finite elements, with boundary conditions that change according to the dislocation structure and the external load.

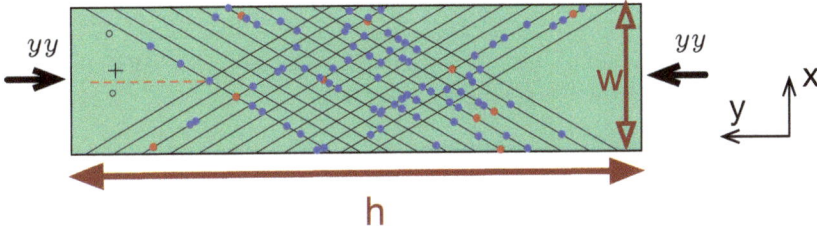

Figure 2. Schematic of the discrete dislocation model in this study. Red dots stand for dislocation sources and blue dots represent dislocation obstacles. Black lines stand for slip planes. Slip planes that cross loading edges are not considered to avoid possible numerical issues caused by dislocations pinned at loading edges. This simplification/assumption does not alter the main mechanism of dislocation interactions. The slip system angle are also indicated in the figure separated by the red dashed line.

Slip planes are spaced at $10b$, where b is the Burgers vector magnitude of 0.25 nm. We do not consider slip planes that cross the loading boundaries (see Figure 2) to avoid numerical difficulties induced by dislocations hitting the boundaries. Such assumptions will not alter the plasticity mechanism observed in the sample since the effective slip area is 85% of the sample geometry. The crystal is initially stress and mobile-dislocation free. This stands for a well-annealed sample, yet with pinned dislocation segments left that can act either as dislocation sources or as obstacles. A dislocation source mimics the Frank–Read source in two dimensions [36]. Point obstacles are included to account for the effect of blocked slip caused by precipitates and forest dislocations on out-of-plane slip systems that are not explicitly described. Stress caused by the obstacles is not considered in the model. The strength of the obstacles τ_{obs} is taken to be 150 MPa with 20% standard deviation. Obstacles are randomly distributed over the slip planes with a density that is eight times the source density [35], and a dislocation stays pinned until its Peach–Koehler force exceeds the obstacle-dependent value $\tau_{obs}b$.

A dipole will be generated from a source when the resolved shear stress τ at the source location is sufficiently high (satisfying the condition $\tau > \tau_{nuc}$) for a sufficiently long time (t_{nuc}). The sources are randomly distributed over slip planes at a density ρ_{nuc} (60 μm^{-2}), while their strength is selected randomly from a Gaussian distribution with mean value $\bar{\tau}_{nuc} = 50$ MPa and standard deviation

10 MPa. Once the source is activated, a dipole is generated and put at a distance L_{nuc}. The initial distance between the two dislocations in the dipole is

$$L_{nuc} = \frac{E}{4\pi(1-\nu^2)} \frac{b}{\tau_{nuc}},\qquad(3)$$

at which the shear stress of one dislocation acting on the other is balanced by the local shear stress which equals τ_{nuc}. After a dislocation is nucleated, it can either exit the sample through the traction-free surface, annihilate with a dislocation of opposite sign when their mutual distance is less than $6b$ or become pinned at an obstacle when the dislocation moves to the obstacle site.

Glide is governed by the component of the Peach–Koehler force in the slip direction. For the I-th dislocation, this force is given by

$$f^{(I)} = \mathbf{n}^{(I)} \cdot \left(\hat{\sigma} + \sum_{J \neq I} \tilde{\sigma}^{(J)} \right) \cdot \mathbf{b}^{(I)},\qquad(4)$$

where $\mathbf{n}^{(I)}$ is the slip plane normal and $\mathbf{b}^{(I)}$ is the Burgers vector of dislocation I. The Peach–Koehler force (Equation (4)) includes the stress contribution from all other dislocations in the system (sum of ~ fields) and effective stress (^), considering the external loading and correction fields of the superposition method. Dislocations follow over-damped dynamics, therefore they are driven by their Peach–Koehler forces, and the instantaneous velocity of the I-th dislocation is

$$v^{(I)} = \frac{f^{(I)}}{B},\qquad(5)$$

where B is the drag coefficient. In this paper, its value is taken as $B = 10^{-4}$ Pa·s, which is representative of aluminum, with $E = 70$ GPa and $\nu = 0.33$.

Simulations are carried out in an incremental manner, with a time step that is 20 smaller than the nucleation time $t_{nuc} = 10$ ns. At the beginning of every time increment, nucleation, annihilation, pinning at and release from obstacle sites are checked. After updating the dislocation structure, new stress fields in the sample are determined, using the finite element method [36]. The loading mode is set up to be strain rate controlled at 10^4/s. We primarily focus on multi-slip loading (two active slip systems oriented at $\pm 30°$ relative to the loading direction), and compare our results with single-slip systems, with slip orientation $30°$ relative to the loading direction.

3. The Mechanical Response of Finite Small Volumes in Multi-Slip Conditions

First, we investigate the behavior of samples for fixed total deformation strain (1%) in both annealed and mobile-dislocation-rich samples for multi-slip loading conditions. Pre-existing dislocation microstructures in mobile-dislocation-rich samples are altered through the prior deformation of annealed samples at increasing total strain levels (1%, 5%, 10%), as shown in Figure 3 (example of 10% loading history). If uniaxial compression of "annealed" samples (only dislocation sources initially present) is carried out (cf. Figure 3a), then we find a yield strength size effect and stochastic plastic behavior (cf. Figure 3b) that are qualitatively consistent with experimental findings for uniaxial crystal compression of nanopillars [7,28]. When the developed microstructures at certain loading strain are unloaded to zero stress, a stable dislocation structure of a mobile-dislocation-rich sample forms, as shown in Figure 3c. The mobile-dislocation-rich sample is then loaded to 1% strain as shown in Figure 3d.

The direct comparison of mechanical responses for small finite volumes of different sizes (different colors) and microstructures (dashed vs. solid) is shown in Figure 4. The statistical averages of stress–strain curves based on 50 samples are plotted in Figure 4a. Loading of pre-existing dislocation ensembles to 1% total strain leads to nonlinearity, i.e., a smooth and nonlinear response prior to reaching the flow stress. The average curvature drastically decreases as the sample size decreases i.e., a longer continuous transition from elastic to perfect plastic, in contrast to the expected discontinuous yielding of annealed structures (dashed lines).

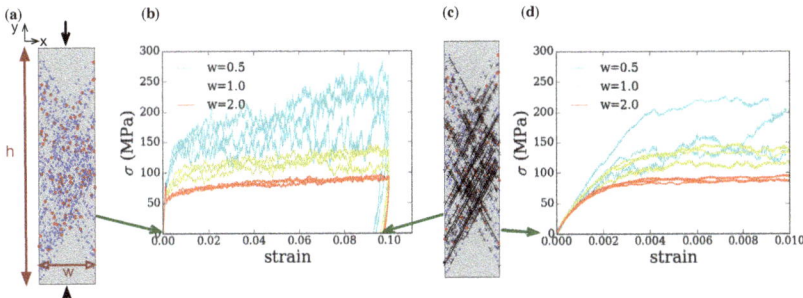

Figure 3. Annealed vs. Mobile-Dislocation-Rich dislocation microstructures and uniaxial compression in small finite volumes. (a) Initially annealed dislocation structure: large (red) dots stand for dislocation sources, small (blue) dots stand for obstacles, and two slip systems are used. The pillar has aspect ratio $h/w = 4$; (b) Examples of stress–strain curves of loading-unloading process for different sizes w, 20 realizations each; for each w three of them are shown. (c) Dislocation structure after unloading (one representative structure is shown for $w = 2$ µm), the average dislocation densities ($10^{14}/m^2$) for decreasing w are 11, 10.6, 8.6. (d) Examples of reloading of the pre-existing dislocation microstructure.

The observed nonlinear behavior is evidently related to the yield strength size effect in small volumes: while the ensemble average of the yield strength increases as $w \to 0$ (see Figure 4b) for either annealed or loaded microstructures, the yield strength *distribution* (see Figure 4b) becomes drastically wider with system-size for loaded dislocation configurations, in a qualitative agreement with nanopillar compression phenomenology [37]. By comparing Figure 4a,b, one may notice that the yield stress distribution disparity mirrors the system-size dependence of the anelastic (nonlinear) average behavior. The same exponent that controls the yield strength size effect ($\sigma_Y \sim w^{-\alpha}$ with $\alpha \simeq 0.65$) [35] is the one that controls the nonlinearity of the average stress–strain behavior (not shown). This finding is consistent with recent observations (e.g., see Ref. [16]—Appendix Figure S4d).

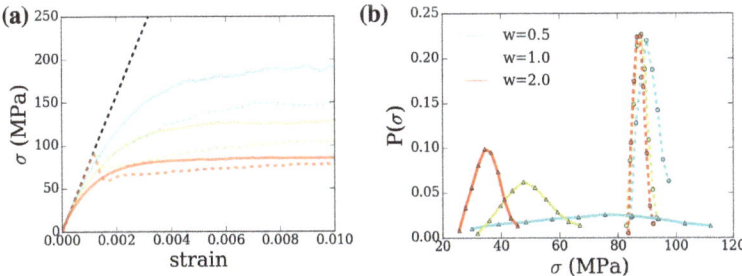

Figure 4. "Annealed" vs. "Mobile-Dislocation-Rich" Ensemble Averages in Multi-Slip Conditions. (a) Average stress–strain curves of different w (50 realizations each). The black dashed line indicates the expected elastic behavior due to the material's elastic modulus. Colored dashed lines are the average stress–strain curves up to 1% strain shown in Figure 3b when loading the annealed microstructure. Solid colored lines are the average stress–strain curves for pre-loaded 10% samples. (b) Yielding (defined as 0.1% plastic strain) distribution for different w. The line type follows (a).

4. Dislocation Pair Correlations and Single-Slip vs. Multi-Slip Loading Conditions

The effects seen in double-slip loading conditions are not generic. The observed nonlinearity depends on the number of slip systems activated, so we compare results produced in single-slip (oriented at $-30°$, see Figure 2) and multi-slip loading conditions. The nonlinearity becomes clear when the stress–strain curve is reconstructed by defining $\sigma_r = \sigma - \sigma_f$, where σ_f is the flow stress prior to unloading. It is seen in Figure 5, where σ_r versus plastic strain ϵ^p, is plotted that the nonlinearity has a dependence on the sample size for double-slip loading. In contrast, single-slip loading shows no clear size dependent nonlinearity for mobile-dislocation-rich samples (see Figure 5a inset). This apparent discrepancy between single-slip and multi-slip loading indicates a possible connection of this size effect to certain spatial features of dislocation structures that are favorably formed under double slip conditions.

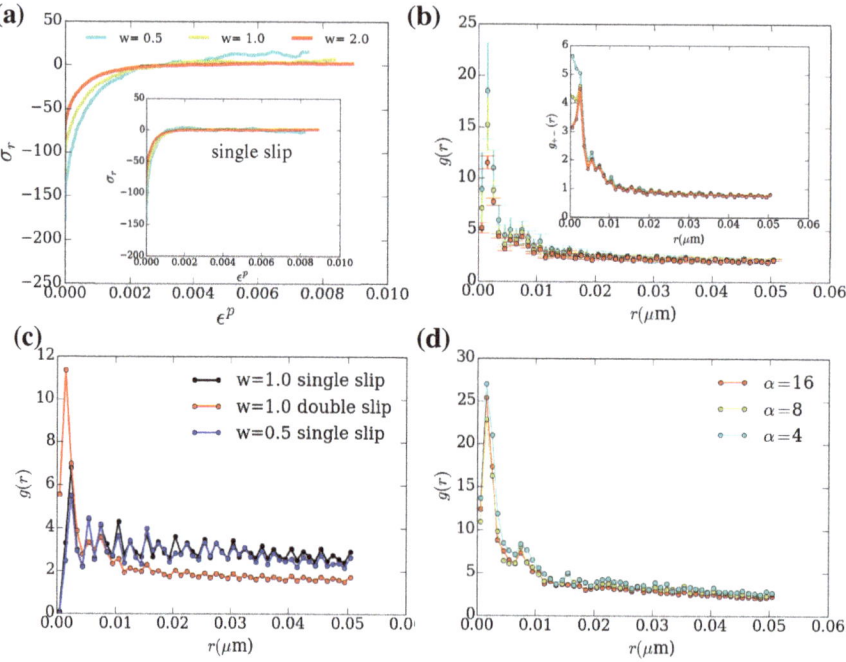

Figure 5. Single-Slip vs. Multi-Slip Loading Conditions and Structural Correlations. (**a**) Reconstructed stress–strain curves for different w with $\sigma_r = \sigma - \sigma_f$ where σ_f is the flow stress prior to unloading. The inset is the reconstructed stress–strain curves when single slip system is used in the modeling. (**b**) Pair correlation $g(r)$ of dislocation structure obtained by unloading. The inset shows $g_{+-}(r)$ correlation. The color of each curve corresponds to w show in panel (**a**). (**c**) Pair correlation $g(r)$ of dislocation structure in double and single slip systems. (**d**) Pair correlation $g(r)$ of dislocation structure in samples of different aspect ratio α and fixed $w = 1.0$.

Spatial features of dislocation structures may be extracted by: (i) the study of pair correlation functions $g_{ss'}(r)$ where $s, s' \in \{+, -\}$, as well as (ii) the sign-insensitive correlation function $g(r)$, with $r = \sqrt{(x - x')^2 + (y - y')^2}$ for two dislocations located at $\mathbf{r} = (x, y)$ and $\mathbf{r}' = (x', y')$. Figure 5b shows $g(r)$ for different w, averaged over 20 realizations. A structural peak forms in $g(r)$ at small distances (~ 2 nm, with the slip spacing being 2.5 nm), which signifies the formation of dislocation dipoles. The clustered dislocations are not pile ups (at single slip planes) as we confirmed. The scatter of the pair correlation (errorbar shown in Figure 5b) increases with decreasing sample size, consistently with the variability of yield stress shown in Figure 4b. The origin of the pairs can be traced in the dynamical behavior of the model: Dislocations from different slip systems may mutually approach at a very short distance without annihilation, at the intersection of their respective slip planes. There, a stable structure can be formed by dislocations of opposite signs (see Figure 3c for example). The inset shows the behavior of the average $g_{+-}(r)$: pairs of dislocations with opposite signs are clustered at distances smaller than 3 nm with the peak of $g_{+-}(r)$ being higher as $w \to 0$. Dislocation pairs of opposite-signed dislocations may be viewed as a toy model of dislocation junctions [7], even though such analogies should be considered with care. In single slip loading, as shown in Figure 5c, the peak of the pair correlation function appears exactly at the slip plane spacing 2.5 nm, larger than that in the double slip system. For consistency purposes, we also checked analogous results in samples of different aspect ratios, one of them being shown in Figure 5d for $w = 1$ μm, and no clear difference is found, thus we conclude that $\alpha = 4$ is adequate for the purposes of this study.

The very formation of bound dislocation dipoles may not necessarily imply any size dependence of the nonlinear mechanical response [3,6,7,38]. However, the origin of the correlated size-dependent response is indeed tracked down to the stress-field imposed by these inter-slip dislocation pairs. Namely, a single edge dislocation displays a long-range resolved shear stress that has *stability lines* at 45° angle with respect to the slip system angle, and an opposite-signed dislocation can combine to form a bound pair at a nearby slip plane. The inter-slip bound dislocation pairs, discussed in this work, apply equally long-range dislocation stress, as the one originating in a single dislocation. The dislocation pair can be regarded as a *super-dislocation* where the resolved shear stress along the −30° slip system is plotted in Figure 6a. There are multiple stability lines that can lead to the *kinetic* formation of stable but weak pairs, ultimately leading to a size-dependent correlated response. For each such super-dislocation, the shear stress sign-changing locations are shown with green lines; along such lines, it is probable to stochastically form a wall of such super-dislocation dipoles.

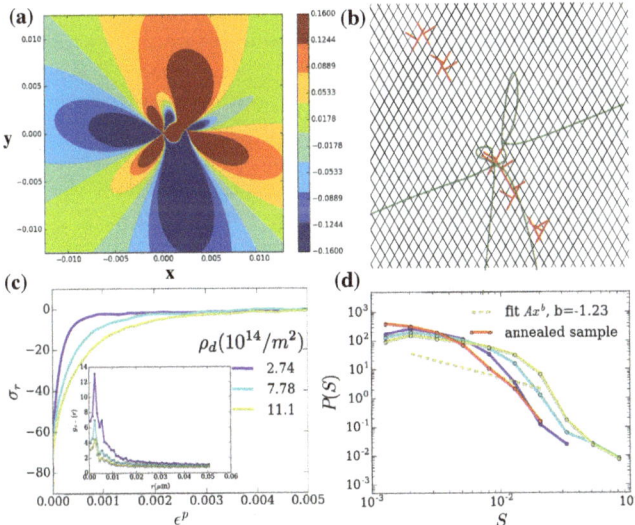

Figure 6. Spatial correlations and the origin of the stress-response dependence on initial dislocation density. (a) Resolved shear stress (unit GPa) along −30° slip system from a +− dislocation pair that typically form in the system, functioning as super-dislocations. (b) Zoom-in of typical dislocation structure formed after unloading for $w = 2$ μm. (c) Reconstructed stress vs. plastic strain for structures with different dislocation densities. The inset shows $g_{+-}(r)$ for different dislocation densities. (d) Statistics of plastic events during reloading. Plastic event S is defined as $S = \sum_{i \in \text{eventsteps}} \delta\sigma_i / \sigma_{\max}$. The red line stands for well-annealed system (see Figure 3a left) loading to 1% strain.

Naturally, the formation of the identified dipoles and the associated patterns should become more probable as the dislocation density increases for the same system size (or as the sample size increases for the same dislocation density). For this purpose, we investigate the effect of different dislocation densities (for $w = 2$ μm) through creating dislocation ensembles by unloading at different strains (1%, 5% and 10%). We consider dislocation densities ($10^{14}/m^2$) that are 2.74, 7.78, 11.1 (see Figure 6c). It is seen that larger initial dislocation density leads to a more pronounced nonlinearity. The $g_{+-}(r)$ is shown in the inset, which signifies that the smaller dislocation density has a higher peak. Our model is benchmarked with experimental data [35], i.e., and it predicts a realistic strengthening size effect and dislocation avalanche statistics in FCC crystals even though at a much higher strain rate than experiments. Assuming a sole dislocation density effect on the strength, we may estimate that the pre-existing dislocation density of the samples in Ref. [16,17] to be $10^{13}/m^2$.

The evolution of the *average* inelasticity may be tracked through the statistics of abrupt events that caused it. The event size statistics is shown in Figure 6d. Event S is the normalized stress drop defined as $\sum_{i \in eventsteps} \delta\sigma_i / \sigma_{max}$ where $\delta\sigma_i$ is the stress drop and σ_{max} is the maximum stress in single realization. It can be clearly seen that the increase of pre-existing dislocation density and inelasticity leads to a power-law behaving ensemble with larger cutoff and decay exponent ~1.23. For very low pre-existing dislocation density, where crystal plasticity is dominated by dislocation nucleation, one can see that the power-law behavior is almost invisible.

5. Conclusions

This result is in accordance with a wealth of prior work [7,39–45] that have pointed that critical avalanche dynamics requires pre-existing random or "glassy" dislocation microstructures. However, the current work represents a pioneering effort to identify the precise origin of such random structures in small scales. The possible distinction of this work is the fact that realistic dislocation microstructure formation, contrary to a purely random dislocation microstructure, may lead to clear power-law abrupt event statistics and associated effects.

In summary, we identified and studied a nonlinearity in the stress–strain *initial-condition ensemble average* response during uniaxial compression of small finite volumes. This nonlinear effect is an outcome of small finite-volume avalanche responses [7] and its presence may challenge any possible correspondence between large-scale mechanical response and ensemble averages of small finite volumes (see Figure 1). We find that such correspondence is plausible and sensible for single-slip loading conditions and sample widths down to 500 nm, but not for multi-slip loading conditions with sample widths up to 2 μm. We track the very origin of this effect in the structural features of the emerging dislocation structures and the formation of bound dislocation dipoles.

This dipole formation resembles dislocation junction formation in more detailed models of dislocation dynamics [7]. We may consider a typical macroscopic phenomenological power law strain hardening relation to model this effect in continuum plasticity modeling [1], by stating that the post-yield stress, $\sigma = K\epsilon^n$ with K the strength coefficient and n the hardening exponent. We find that n (which is defined as $\frac{\log \sigma}{\log \epsilon}$) is a function of the pre-existing dislocation density ρ, leading to the constitutive relation $n = 1 - (644 - 35.35\rho^*)\epsilon$, where $\rho^* = \rho/\rho_0$ with ρ_0 being $10^{14}/m^2$. Thus, our explicit discrete dislocation model study of uniaxial compression in small finite volumes demonstrates that ubiquitous abrupt plastic events result into a dislocation density dependent nonlinear dependence.

Together with strength size effects, the identified nonlinearity challenges any attempts for "ensemble averaging" of small-volume responses into forming a representative volume element average. The density dependence can be traced to the pattern formation of microscopic dislocation dipoles, which are not easily formed in single-slip loading conditions. In this way, multi-slip loading conditions are possibly key components to unveiling the role of critical, power-law abrupt events' for phenomenological crystal plasticity.

Author Contributions: H.S. and S.P. designed the study; H.S. carried out simulations. H.S. and S.P. analyzed the data and wrote the manuscript. All authors reviewed the final manuscript.

Funding: This research was funded by the National Science Foundation under award number 1709568.

Acknowledgments: We would like to thank X. Ni and E. Van der Giessen for inspiring discussions. We also acknowledge the use of the High Performance Computing System (Spruce Knob) of West Virginia University.

Conflicts of Interest: The authors declare no conflict of interest. The funders had no role in the design of the study; in the collection, analyses, or interpretation of data; in the writing of the manuscript, or in the decision to publish the results.

References

1. Needleman, A. Postbifurcation behavior and imperfection sensitivity of elastic-plastic circular plates. *Int. J. Mech. Sci.* **1975**, *17*, 1–13. [CrossRef]
2. Asaro, R.; Lubarda, V. *Mechanics of Solids and Materials*; Cambridge University Press: Cambridge, UK, 2006.
3. Uchic, M.D.; Shade, P.A.; Dimiduk, D.M. Plasticity of micrometer-scale single crystals in compression. *Annu. Rev. Mater. Res.* **2009**, *39*, 361–386. [CrossRef]
4. Dimiduk, D.M.; Uchic, M.D.; Rao, S.; Woodward, C.; Parthasarathy, T. Overview of experiments on microcrystal plasticity in FCC-derivative materials: Selected challenges for modelling and simulation of plasticity. *Model. Simul. Mater. Sci. Eng.* **2007**, *15*, 135. [CrossRef]
5. Uchic, M.D.; Dimiduk, D.M.; Florando, J.; Nix, W. Exploring specimen size effects in plastic deformation of Ni_3 (Al, Ta). *MRS Proc.* **2002**, *753*. [CrossRef]
6. Greer, J.R.; De Hosson, J.T.M. Plasticity in small-sized metallic systems: Intrinsic versus extrinsic size effect. *Prog. Mater. Sci.* **2011**, *56*, 654–724. [CrossRef]
7. Papanikolaou, S.; Cui, Y.; Ghoniem, N. Avalanches and plastic flow in crystal plasticity: An overview. *Model. Simul. Mater. Sci. Eng.* **2017**, *26*, 013001. [CrossRef]
8. Cui, Y.; Lin, P.; Liu, Z.; Zhuang, Z. Theoretical and numerical investigations of single arm dislocation source controlled plastic flow in FCC micropillars. *Int. J. Plast.* **2014**, *55*, 279–292. [CrossRef]
9. Jennings, A.T.; Li, J.; Greer, J.R. Emergence of strain-rate sensitivity in Cu nanopillars: Transition from dislocation multiplication to dislocation nucleation. *Acta Mater.* **2011**, *59*, 5627–5637. [CrossRef]
10. Agnihotri, P.K.; Van der Giessen, E. On the rate sensitivity in discrete dislocation plasticity. *Mech. Mater.* **2015**, *90*, 37–46. [CrossRef]
11. Xiang, Y.; Vlassak, J. Bauschinger and size effects in thin-film plasticity. *Acta Mater.* **2006**, *54*, 5449–5460. [CrossRef]
12. Nicola, L.; Xiang, Y.; Vlassak, J.; Van der Giessen, E.; Needleman, A. Plastic deformation of freestanding thin films: Experiments and modeling. *J. Mech. Phys. Solids* **2006**, *54*, 2089–2110. [CrossRef]
13. Shishvan, S.S.; Van der Giessen, E. Distribution of dislocation source length and the size dependent yield strength in freestanding thin films. *J. Mech. Phys. Solids* **2010**, *58*, 678–695. [CrossRef]
14. Chan, P.Y.; Tsekenis, G.; Dantzig, J.; Dahmen, K.A.; Goldenfeld, N. Plasticity and dislocation dynamics in a phase field crystal model. *Phys. Rev. Lett.* **2010**, *105*, 015502. [CrossRef]
15. Zhang, P.; Salman, O.U.; Zhang, J.Y.; Liu, G.; Weiss, J.; Truskinovsky, L.; Sun, J. Taming intermittent plasticity at small scales. *Acta Mater.* **2017**, *128*, 351–364. [CrossRef]
16. Ni, X.; Zhang, H.; Liarte, D.B.; McFaul, L.W.; Dahmen, K.A.; Sethna, J.P.; Greer, J.R. Yield precursor dislocation avalanches in small crystals: The irreversibility transition. *arXiv* **2018**, arXiv:1802.04040.
17. Ni, X.; Papanikolaou, S.; Vajente, G.; Adhikari, R.X.; Greer, J.R. Probing microplasticity in small-scale fcc crystals via dynamic mechanical analysis. *Phys. Rev. Lett.* **2017**, *118*, 155501. [CrossRef] [PubMed]
18. Zener, C. *Elasticity and Anelasticity of Metals*; University of Chicago Press: Chicago, IL, USA, 1948.
19. Cleveland, R.; Ghosh, A. Inelastic effects on springback in metals. *Int. J. Plast.* **2002**, *18*, 769–785. [CrossRef]
20. Kim, H.; Kim, C.; Barlat, F.; Pavlina, E.; Lee, M.G. Nonlinear elastic behaviors of low and high strength steels in unloading and reloading. *Mater. Sci. Eng. A* **2013**, *562*, 161–171. [CrossRef]
21. Ghosh, A. A physically-based constitutive model for metal deformation. *Acta Metall.* **1980**, *28*, 1443–1465. [CrossRef]
22. Perez, R.; Benito, J.; Prado, J. Study of the inelastic response of TRIP steels after plastic deformation. *ISIJ Int.* **2005**, *45*, 1925–1933. [CrossRef]

23. van Liempt, P.; Sietsma, J. A physically based yield criterion I. Determination of the yield stress based on analysis of pre-yield dislocation behaviour. *Mater. Sci. Eng. A* **2016**, *662*, 80–87. [CrossRef]
24. Arechabaleta, Z.; van Liempt, P.; Sietsma, J. Quantification of dislocation structures from anelastic deformation behaviour. *Acta Mater.* **2016**, *115*, 314–323. [CrossRef]
25. Greer, J.R.; Oliver, W.C.; Nix, W.D. Size dependence of mechanical properties of gold at the micron scale in the absence of strain gradients. *Acta Mater.* **2005**, *53*, 1821–1830. [CrossRef]
26. Greer, J.R.; Nix, W.D. Nanoscale gold pillars strengthened through dislocation starvation. *Phys. Rev. B* **2006**, *73*, 245410. [CrossRef]
27. Shan, Z.; Mishra, R.K.; Asif, S.S.; Warren, O.L.; Minor, A.M. Mechanical annealing and source-limited deformation in submicrometre-diameter Ni crystals. *Nat. Mater.* **2008**, *7*, 115–119. [CrossRef] [PubMed]
28. Sethna, J.P.; Bierbaum, M.K.; Dahmen, K.A.; Goodrich, C.P.; Greer, J.R.; Hayden, L.X.; Kent-Dobias, J.P.; Lee, E.D.; Liarte, D.B.; Ni, X.; et al. Deformation of crystals: Connections with statistical physics. *Annu. Rev. Mater. Res.* **2017**, *47*, 217–246. [CrossRef]
29. Anderson, P.M.; Hirth, J.P.; Lothe, J. *Theory of Dislocations*; Cambridge University Press: Cambridge, UK, 1982.
30. Kleemola, H.; Nieminen, M. On the strain-hardening parameters of metals. *Metall. Trans.* **1974**, *5*, 1863–1866. [CrossRef]
31. Ghosh, A. The influence of strain hardening and strain-rate sensitivity on sheet metal forming. *J. Eng. Mater. Technol.* **1977**, *99*, 264–274. [CrossRef]
32. Peirce, D.; Asaro, R.J.; Needleman, A. Material rate dependence and localized deformation in crystalline solids. *Acta Metall.* **1983**, *31*, 1951–1976. [CrossRef]
33. Ghosh, A. Tensile instability and necking in materials with strain hardening and strain-rate hardening. *Acta Metall.* **1977**, *25*, 1413–1424. [CrossRef]
34. Fan, Z.; Mingzhi, H.; Deke, S. The relationship between the strain-hardening exponent n and the microstructure of metals. *Mater. Sci. Eng. A* **1989**, *122*, 211–213. [CrossRef]
35. Papanikolaou, S.; Song, H.; Van der Giessen, E. Obstacles and sources in dislocation dynamics: Strengthening and statistics of abrupt plastic events in nanopillar compression. *J. Mech. Phys. Solids* **2017**, *102*, 17–29. [CrossRef]
36. Van der Giessen, E.; Needleman, A. Discrete dislocation plasticity: A simple planar model. *Model. Simul. Mater. Sci. Eng.* **1995**, *3*, 689. [CrossRef]
37. Papanikolaou, S. Learning local, quenched disorder in plasticity and other crackling noise phenomena. *arXiv* **2018**, arXiv:1803.03603.
38. Chaikin, P.M.; Lubensky, T.C.; Witten, T.A. *Principles of Condensed Matter Physics*; Cambridge University Press: Cambridge, UK, 2000.
39. Ispánovity, P.D.; Laurson, L.; Zaiser, M.; Groma, I.; Zapperi, S.; Alava, M.J. Avalanches in 2D dislocation systems: Plastic yielding is not depinning. *Phys. Rev. Lett.* **2014**, *112*, 235501. [CrossRef] [PubMed]
40. Zaiser, M. Scale invariance in plastic flow of crystalline solids. *Adv. Phys.* **2006**, *55*, 185–245. [CrossRef]
41. Tsekenis, G.; Goldenfeld, N.; Dahmen, K.A. Dislocations jam at any density. *Phys. Rev. Lett.* **2011**, *106*, 105501. [CrossRef] [PubMed]
42. Zaiser, M.; Aifantis, E.C. Randomness and slip avalanches in gradient plasticity. *Int. J. Plast.* **2006**, *22*, 1432–1455. [CrossRef]
43. Papanikolaou, S.; Bohn, F.; Sommer, R.; Durin, G.; Zapperi, S.; Sethna, J. Universality beyond power laws and the average avalanche shape. *Nat. Phys.* **2011**, *7*, 316–320. [CrossRef]
44. Papanikolaou, S.; Dimiduk, D.; Choi, W.; Sethna, J.; Uchic, M.; Woodward, C.; Zapperi, S. Quasi-periodic events in crystal plasticity and the self-organized avalanche oscillator. *Nature* **2012**, *490*, 517–521. [CrossRef]
45. Miguel, M.C.; Vespignani, A.; Zapperi, S.; Weiss, J.; Grasso, J.R. Intermittent dislocation flow in viscoplastic deformation. *Nature* **2001**, *410*, 667–671. [CrossRef] [PubMed]

© 2019 by the authors. Licensee MDPI, Basel, Switzerland. This article is an open access article distributed under the terms and conditions of the Creative Commons Attribution (CC BY) license (http://creativecommons.org/licenses/by/4.0/).

Article

Influence of Size on the Fractal Dimension of Dislocation Microstructure

Yinan Cui * and Nasr Ghoniem

Department of Mechanical and Aerospace Engineering, University of California Los Angeles, Los Angeles, CA 90095, USA; nghoniem@gmail.com
* Correspondence: cuiyinan@g.ucla.edu; Tel.: +1-310-825-4866

Received: 29 March 2019; Accepted: 20 April 2019; Published: 25 April 2019

Abstract: Three-dimensional (3D) discrete dislocation dynamics simulations are used to analyze the size effect on the fractal dimension of two-dimensional (2D) and 3D dislocation microstructure. 2D dislocation structures are analyzed first, and the calculated fractal dimension (n_2) is found to be consistent with experimental results gleaned from transmission electron microscopy images. The value of n_2 is found to be close to unity for sizes smaller than 300 nm, and increases to a saturation value of ≈1.8 for sizes above approximately 10 microns. It is discovered that reducing the sample size leads to a decrease in the fractal dimension because of the decrease in the likelihood of forming strong tangles at small scales. Dislocation ensembles are found to exist in a more isolated way at the nano- and micro-scales. Fractal analysis is carried out on 3D dislocation structures and the 3D fractal dimension (n_3) is determined. The analysis here shows that (n_3) is significantly smaller than ($n_2 + 1$) of 2D projected dislocations in all considered sizes.

Keywords: dislocation microstructure; fractal analysis; size effect

1. Introduction

Dislocations are the main carriers of plastic deformation in crystals. Within the framework of crystal plasticity theory, dislocations are generally described by their density. Key mechanical properties were found to correlate with the concept of density, for example the Taylor hardening law, which states that the critical resolved shear stress is proportional to the square root of the dislocation density. To account for size effects on plastic deformation, further refinements have led to the distinction between geometrically necessary and statistically stored components. This distinction allows consideration of the role of dislocation accumulation in accommodating an imposed deformation gradient and the development of strain gradient plasticity theory. The spatial consideration of gradients in the dislocation density resulted in successful interpretation of the size effect in micro-bending and micro-torsion tests. Going beyond this mean-field description through the concept of dislocation density, the complex spatial features of the dislocation structure have also attracted considerable interest [1–9]. The formation of planar dislocation arrays is known to be a prelude to micro shear banding, while the entanglement of dislocations can lead to greater work-hardening, and the size of dislocation cells may be a reflection of creep strength [10–12]. Therefore, effective description of the dislocation structure is important for building the relationship between microstructure and the mechanical behavior. Some typically observed dislocation pattern morphology includes ladder, labyrinth, wall, and cell structures, which depend on the material, loading condition, and temperature, etc. Complete quantitative characterization of the dislocation structure is very difficult because the large number of parameters that may be necessary [13]. Investigation of the common features from a statistical perspective leads to a practical way to describe complicated dislocation structures in a simple way. An interesting finding is the fractal nature of dislocation structures [13–16] as a measure of structural complexity and spatial packing.

Fractals generally reflect statistical self-similarity. Specifically, spatial features look similar at any scale. This implies scale-free physics. In addition, geometrical structures have features at all length scales. One typical example to gain an intuitive idea is the measurement of the length of a coastline [17]. When one uses a very large ruler to measure the length, one obtains a smaller estimate of the length than using a fine ruler, which can capture more details on the smaller scale [18]. The measured length L is expressed by the number of scale (ruler) units $N(\Delta x)$. Here, each unit has the length of the ruler size Δx. A power law relation is observed between N and the ruler size Δx as $N \propto \Delta x^{-n}$. The absolute value of the corresponding power law exponent is defined as a fractal dimension n. n can be non-integer values. This idea can be extended to two-dimensional (2D) and three-dimensional (3D) systems, by tuning the dimensionality of the ruler. For ordinary geometric shapes, the theoretical fractal dimension is equal to its topological dimension. For a fractal geometry, the fractal dimension exceeds its topological dimension. The fractal dimension quantifies the complexity as a ratio of the change in detail to the change in scale.

The plastic deformation of a material is a highly complex spatio-temporal phenomenon. The complexity of the underlying dynamics is mainly associated with the nonlinear evolution of collective dislocations. In the temporal scale, it is manifested as the emergence of strain burst and dislocation avalanches [8]. In the spatial scale, this leads to a non-uniform, non-isotropic, and non-random dislocation distributions [19]. Regular, periodic, or nearly periodic dislocation patterns, such as persistent slip bands with a well-defined pattern wavelength [13], are not fractal patterning. However, the widely observed dislocation cellular structure generally exhibits a clear fractal geometry, which implies scale invariance of the spatial arrangement of dislocations at a given deformation state. The fractal dimension measures the space-filling capacity and the complexity of a dislocation pattern [7]. As summarized in Table 1, the fractal dimension of cellular dislocation patterns depends on strain value [20] and stress level. During the initial stages of deformation, the dislocation structure evolves significantly before a relatively stable dislocation structure is formed. The fractal dimension gradually increases at first, and then reaches a relatively stable value with increasing strain [16,20]. When the strain is close to the onset of necking, the fractal dimension of the bulk dislocation structure starts to gradually drop, because the structure becomes progressively more ordered as fracture is approached [20]. This leads to the possibility of linking the variation of the fractal dimension of the dislocation structure with strain, which may contribute to the development of a constitutive law. When the external size of the material decreases to several microns, the fractal dimension also shows dependence on the sample size [16].

Table 1. Fractal dimension of dislocation cellular structures estimated through experiments or theoretical models.

Fractal Dimension n	Remarks	Reference
1.64~1.79	TEM images of [100]-oriented Cu single crystal, n depends on stress.	[13,21,22]
1.371~1.695	STEM images of [011]-oriented Cu, n depends on sizes and strain.	[16]
1.33	2D dislocation glide through obstacles.	[23]
0.9~1.8	Phase-field simulations, n initially increases with applied stress.	[24]
1.5	Using 2D continuum model for mesoscale plasticity.	[25]
1.87	2D dislocation pattern simulation for FCC single crystals oriented for multiple slip without climb under cyclic loading.	[26]

The fractal feature of the dislocation microstructure is generally studied based on the analysis of transmission electron microscopy (TEM) or scanning transmission electron microscopy (STEM) micrographs [13,16]. An analysis method based on box-counting will be described in detail in Section 2. Other alternative measurement methods are referred to in reference Zaiser et al. [13]. As we know, TEM or STEM images can be considered to be 2D projections of the actual 3D dislocation microstructure. Most of existing theoretical model used to describe the fractal feature of dislocations also mainly focuses on 2D problems, as summarized in Table 1. Then, an interesting question is whether the fractal behavior can be observed for the actual 3D dislocation microstructure? If yes, then a follow-up question would be how the external size influences the fractal behavior of 3D dislocation structure? Till now, reconstruction of the 3D dislocation microstructure through experimental data is still very difficult. Only recently, some efforts of visualizing 3D dislocation structure have been spent by using X-ray tomography [27], using the electron beam tomography method through tilting samples while maintaining proper diffraction condition [28–30], or using scanning electron microscope serial sectioning method [31]. However, the obtained 3D dislocation structure information is still limited and not used to check their fractal behavior yet. On the other hand, discrete dislocation dynamics (DDD) simulations represent a powerful tool to investigate the formation of dislocation structure by considering the short-range and long-range dislocation interactions and external load conditions [32–34]. 2D-DDD reproduces the fractal dislocation structures in multiple slip [15]. It is also found that self-similar dislocation patterns form without dislocation climb, but cellular structures with well-defined characteristic lengths are observed with dislocation climb [26]. To our knowledge, 3D-DDD is not used to investigate the fractal nature of dislocation structures yet due to the significant computational expense when simulating highly tangled dislocation cells. Therefore, it is not clear whether 3D dislocation structures self-organize into fractal features through junction formation, pile-ups, cross-slip events, etc.

Based on the considerations above, the current work aims to answer two questions. The first is whether a 3D dislocation structure exhibits well-defined fractal features. The second is how the sample size influences the fractal behavior of 2D projected dislocation structures and actual 3D space-filling tangles. In Section 2, the investigation method is described. Simulation results and discussions are given in Section 3, while Section 4 summarizes the results of the present investigation.

2. Investigation Method

The computational method of 3D-DDD is employed here as one component of the MoDELib (Mechanics of Defect Evolution Library) software system [35], described in detail in our previous papers [36,37]. In this 3D-DDD approach, curved dislocation lines are discretized into a succession of parametrized segments. Boundary conditions and image forces are considered by coupling with an FEM solution of an elasticity problem using the superposition principle [37]. 3D-DDD simulations of tension tests of Fe micropillars along the [001] direction at 320 K are carried out. Pillar diameter varied in the range 300–1500 nm, with the ratio of height and diameter be equal to 2. To tune the extent of deformation localization, we introduce irradiation defects, with defect density range of 10^{21}–3×10^{22} m^{-3}. Irradiation defects themselves are not considered when discussing the fractal behavior of dislocation structures. More details on the simulation set-up and descriptions are given in [33,38–40].

The fractal behavior is investigated by the box-counting method [13,16]. The basic idea of box-counting method is that the space is discretized into a large amount of non-overlapping small grids. If the dislocation line passes through a specific grid "pixel", it is marked as 1, otherwise, the pixel is marked as 0. For example, for the 2D case shown in Figure 1, the filled pixels are marked as 1, and white pixels are marked as 0. In previous work, such information is mainly obtained by post processing experimental images, which is limited by the image resolution. The calculation here is directly based on the information of dislocation segment positions, which allows for very high spatial resolution, and is also applicable to the 3D case. Please note that because we check the intersection of the dislocation line with the grid pixel, instead of only using the information of the ending points of dislocation segments, the calculation is not sensitive to the discretization of the dislocation lines. In addition, we verified with this method the fractal dimension of a perfect circular dislocation loop, and found its fractal dimension to be 1, which is the same as that of straight dislocation line. This implies that the dislocation curvature itself does not have influence on the calculation of the fractal dimension.

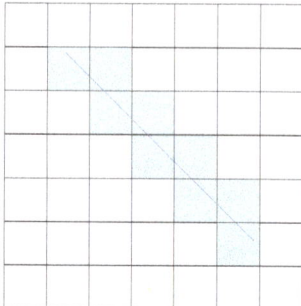

Figure 1. Schematic showing how to convert the dislocation line information into pixel image in binary format. The blue solid line is the dislocation line.

When analyzing the 2D dislocation structure, the 3D dislocation positions are projected along a specific direction ([010] direction is chosen here) to a 2D plane, to compare with the results of TEM images obtained experimentally. Here, we ignore the details of TEM imaging condition. For example, when the dot product of diffraction vector and the burgers vector is zero, the dislocations are invisible. In addition, the considered sizes are close to the allowable thickness of TEM samples, so it is reasonable to consider the projection of all the dislocation structures. The current projected 2D dislocation structure information should still be a very good approximation to the one that will be observed by TEM. After obtaining this information, the number of grids $N(\Delta x)$ those contain dislocations (grids marked as 1) is calculated as a function of the grid size Δx. The fractal dimension can be calculated according to the slope of double-logarithmic plot of $N(\Delta x)$ vs. Δx.

3. Results and Discussion

The simulation results are given in Figure 2 for 2D and 3D dislocation structures in deformed Fe pillars with different sizes, when the applied strain is 1.5%. In the following, we will first analyze the results corresponding to 2D case, and compare with the available experimental results to validate the effectiveness of our calculation and study the size effect.

It is found that when the sample size is as large as 1.5 microns, very good linear behavior is observed in the double-logarithmic plot of $N(\Delta x)$ versus Δx in Figure 2c for 2D box-counting data. A scaling regime with fractional slope of 1.5 extends over almost three orders of magnitude (blue dashed line in Figure 2c. This implies a fractal nature of the dislocation structure, and the corresponding fractal dimension is 1.5. The corresponding dislocation configuration is shown in Figure 2b. One can observe strong dislocation tangles, many small dislocation loops due to dislocation cross-slip and jog formation, and some truncated dislocation lines around the free surface. Previous studies mainly discussed the fractal behavior induced by the multiple dislocation cellular structures. In Figure 2b, even though there are no traditional multiple dislocation cells, the highly tangled dislocation structure exhibits a tree shape, which leads to its fractal behavior. This observation is similar to the experimental image shown in Figure 3 in [16]. To check the sensitivity of the fractal dimensions on the initial dislocation structure and the extent of deformation localization, two other cases are studied, as shown in Figure 3. The results in Figure 3a for #1 is the one shown in Figure 2c. Figure 3a clearly show that for large sample size, the fractal dimension is not sensitive to the initial dislocation structure.

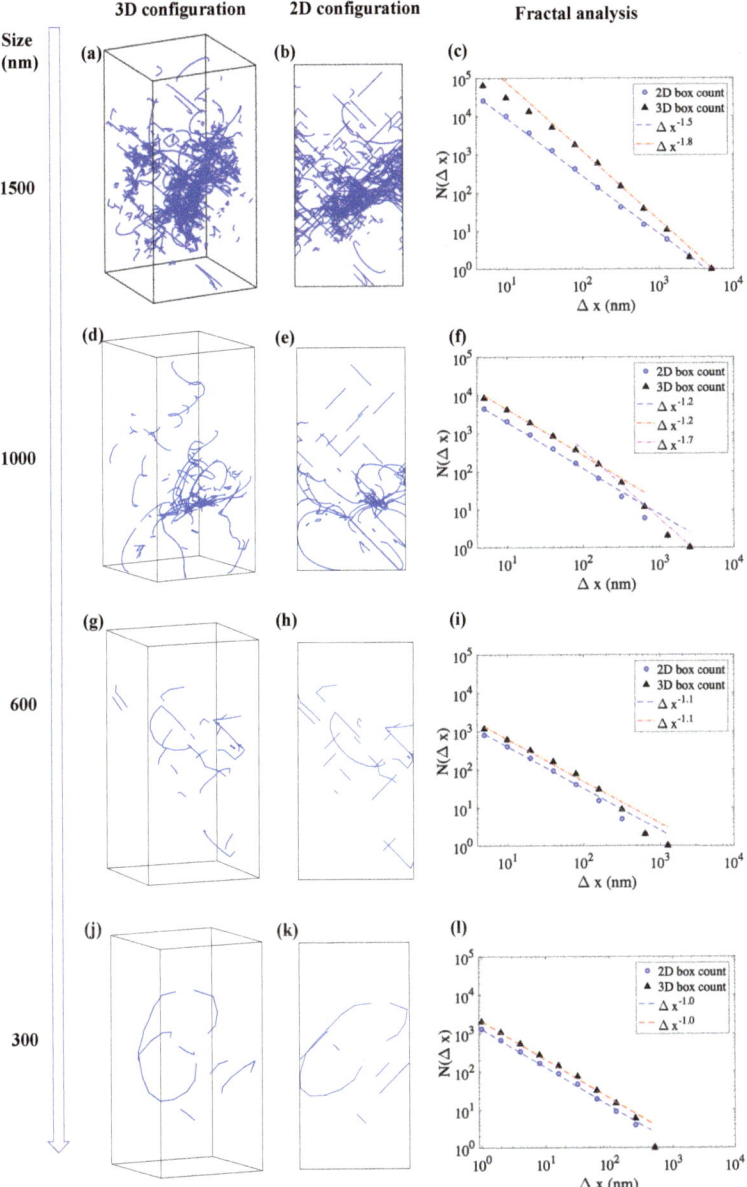

Figure 2. Effect of size and dimension on the fractal analysis of dislocation structure for irradiated Fe pillar with diameter (**a–c**) 1500 nm (**d–f**) 1000 nm, (**g–i**) 600 nm, (**j–l**) 300 nm.

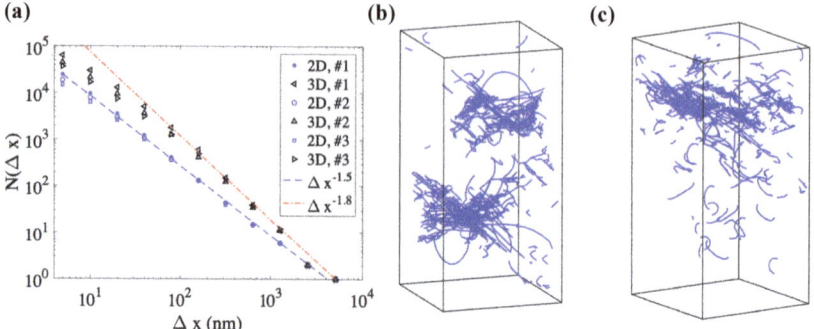

Figure 3. (a) Insensitivity of the fractal dimension of dislocation structure in irradiated Fe pillar with diameter 1500 nm, (b) dislocation configuration #2, (c) dislocation configuration #3.

Figure 2e,f shows the results for an Fe pillar with diameter 1 micron. Compared with Figure 2b, the extent of dislocation tangle is much weaker. The calculated fractal dimension also decreases to 1.2. With further reduction of the sample size, only isolated dislocation lines are observed, as shown in Figure 2h,k. This kind of low dislocation line content is widely observed during compression experiments of nanopillars [41]. This is a result of the ease of dislocation glide out of the crystal through the free surface in such small materials. From Figure 2i,l, even though the double-logarithmic plot of $N(\Delta x)$ versus Δx still shows approximate linear behavior, the calculated exponent is close to unity. When the exponent is close to unity for a 2D image, dislocations are essentially isolated lines, and no fractal behavior exists anymore.

To further compare with the experimental results obtained from TEM images for similar sample sizes, we plot the calculated fractal dimension as a function of sample sizes obtaining from our simulations and recent experimental data in Figure 4. It can be seen that our calculation results agree very well with the experimental results. This clearly demonstrates that reducing the sample size leads to the decrease of the fractal dimension, and the fractal feature of dislocation structures disappears when the sample size is smaller than about 600 nm. The analysis above implies that only when some kind of tangled dislocation structures are observed, the fractal behavior is possible to exist.

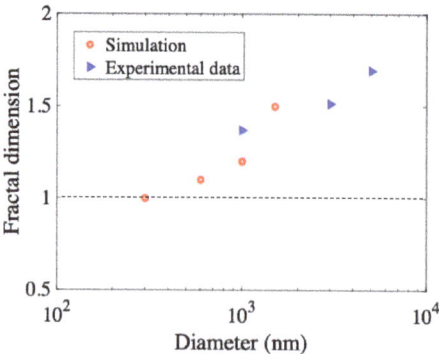

Figure 4. Size effect on the fractal dimension of dislocation microstructure. Experimental data are obtained from [16].

Now we check whether a similar trend exits for 3D dislocation structures as they fill space. As shown in Figure 2c,f,i,l, the scaling regime identified for 3D dislocation structures is shorter than that for the 2D case for the considered sizes. Moreover, the absolute value of the slope for the double-logarithmic plot of $N(\Delta x)$ versus Δx is smaller than 2. Generally, if the fractal dimension is n for 2D case, its corresponding fractal dimension is $n + 1$ for the 3D case, if the structure is extended along the projection direction. This is also widely used to extract the fractal dimension of dislocation structures from 2D results to 3D results [24]. This is true if one uses 2D-DDD, if the calculated fractal dimension is n, its 3D corresponding dimension is $n + 1$, because each point in 2D-DDD correspond to an infinite long straight edge dislocation lines. However, the actual 3D dislocation structure is not the perfect elongation along the projection direction. Therefore, from Figure 2c,f,i,l, one observes that the fractal dimension of 3D dislocation structure is much smaller than one plus the fractal dimension for 2D case, and the fractal behavior in 3D is not as well-defined as the 2D case.

When the sample size is as large as 1.5 microns, most of the 3D box-counting data follows the scaling law with exponent -1.8. Therefore, it is reasonable to say that the fractal dimension of 3D dislocation structure in 1.5 micron diameter pillar is 1.8. From Figures 2a and 3b,c, the highly tangled dislocation configuration is similar to the feature of a tree (see Figure 2a) due to the formation of numerous junctions and jogs, and the occurrence of cross-slip. Therefore, it is natural to think of comparing the calculated fractal dimension with that of a tree. The fractal analysis of the 2D projected and 3D actual tree structure is recently investigated in [42] using box-counting method. The fractal dimension of the 2D projected tree in their studies ranges from 1.69 to 1.94, but the fractal dimension of the corresponding 3D tree ranges from 1.86 to 2.32. This is similar to our observation that the fractal dimension for 3D case is possible to be smaller than 2, and is much smaller than one plus the fractal dimension in 2D. Going back to dislocation configurations, the transition from 2D projected image to 3D dislocation structure is actually similar to the process of pulling up some tangled ropes on the ground. For two dislocation lines, if they do not intersect with each other in 3D, it is still possible to see that their 2D projected lines intersect. Due to 2D projection overlapping, some of the spatial correlations observed in 2D projection image may actually do not exist in 3D. Therefore, it is difficult to observe fractal dimension for 3D case is one plus that of 2D projected dislocations.

When the sample size decreases to be 1 micron, two power law scaling regimes are observed. For the small box size regime, the power law exponent is close to that of 2D dislocations, because the contribution of the isolated dislocation lines is independent of the observation dimensionality. For the large box size regime, the power law exponent is -1.7, which is similar to the fractal dimension of the 1.5 micron diameter pillar. This is contributed by the tangled dislocation structures. For sample size equal to or smaller than 600 nm, most of the scaling regime of the data for 3D dislocation structure has the fractal slope similar to that of 2D projected dislocations. This is due to the lack of highly tangled dislocation structure. Dislocation lines mainly exist in an isolated way.

4. Conclusions

In the current work, the fractal features of dislocation structures in deformed Fe pillars are analyzed through three-dimensional DDD simulations using the box-counting method. It is found that the fractal feature can be well observed for 2D projected dislocation structures, as a result of tangle formation. The fractal dimension decreases with the decrease in the sample size, due to the smaller extent of dislocation entanglement. The results of the current simulations for 2D projected dislocation structures are consistent with experimental results obtained from TEM images. When the sample size is smaller than 300 nm, the fractal feature disappears due to the absence of tangled dislocations. Interestingly, 3D tangled dislocation structures are found to have fractal dimension of 1.7~1.8, which is significantly smaller than expected from 2D analysis alone! This is explained by the fact that 2D projection overlapping leads to spurious dislocation intersections and correlation information that may not be physical. When the external size is smaller than 1000 nm, the fractal slope of 3D box-counting data is similar to that of 2D case, because the fractal dimension of an isolated line is independent of the observation dimensionality.

Further efforts are required to investigate the fractal nature of 3D dislocation structures using other kinds of statistical methods. The evolution of the fractal dimension as a function of strain and dislocation density and its relationship to pattern formation requires more intensive computer simulations to reveal these connections. This requires studies of deformed crystals with larger size and higher dislocation density. The effect of temperature and loading orientation on the fractal dimensions of dislocation structures is also interesting and deserves further studies, especially in BCC crystals where the temperature plays a more significant role.

Author Contributions: Y.C. and G.N. designed the research, discussed the results, and wrote the manuscripts. Y.C. performed the calculation. G.N. gave the constructive suggestions.

Funding: This research was funded by the National Science Foundation, Grant Numbers CMMI-1024353 and CMMI-1727740 at UCLA.

Acknowledgments: Appreciation is expressed to Ronald W. Armstrong and fellow guest editors for the kind invitation to contribute to the Special Issue on "Dislocation Mechanics of Metal Plasticity and Fracturing".

Conflicts of Interest: The authors declare no conflict of interest.

References

1. Kratochvil, J. Dislocation pattern formation in metals. *Rev. Phys. Appl.* **1988**, *23*, 419–429. [CrossRef]
2. Ghoniem, N.M.; Amodeo, R. *Computer Simulaltion of Dislocation Pattern Formation*; Trans Tech Publ.: Zurich, Switzerland, 1988; Volume 3.
3. Sandfeld, S.; Zaiser, M. Pattern formation in a minimal model of continuum dislocation plasticity. *Model. Simul. Mater. Sci. Eng.* **2015**, *23*, 065005. [CrossRef]
4. Groma, I.; Zaiser, M.; Ispánovity, P.D. Dislocation patterning in a two-dimensional continuum theory of dislocations. *Phys. Rev. B* **2016**, *93*, 214110. [CrossRef]
5. Xia, S.; El-Azab, A. Computational modelling of mesoscale dislocation patterning and plastic deformation of single crystals. *Model. Simul. Mater. Sci. Eng.* **2015**, *23*, 055009. [CrossRef]
6. Ngan, A.; Siu, K.; Leung, H.; Cheng, B. Dislocation patterning-meso-scale interactive behavior of dislocations studied through dislocation density-function dynamics. In Proceedings of the 2017 Materials Research Society (MRS) Spring Meeting & Exhibit, Phoenix, AZ, USA, 17–21 April 2017.
7. Li, P.; Zhang, Z. Standing wave effect and fractal structure in dislocation evolution. *Sci. Rep.* **2017**, *7*, 4062. [CrossRef]
8. Papanikolaou, S.; Cui, Y.; Ghoniem, N. Avalanches and plastic flow in crystal plasticity: An overview. *Model. Simul. Mater. Sci. Eng.* **2017**, *26*, 013001. [CrossRef]
9. Cao, L.; Koslowski, M. Effect of microstructural uncertainty on the yield stress of nanocrystalline nickel. *Acta Mater.* **2013**, *61*, 1413–1420. [CrossRef]
10. Lyu, D.; Li, S. A multiscale dislocation pattern dynamics: Towards an atomistic-informed crystal plasticity theory. *J. Mech. Phys. Solids* **2019**, *122*, 613–632. [CrossRef]
11. Irastorza-Landa, A.; Van Swygenhoven, H.; Van Petegem, S.; Grilli, N.; Bollhalder, A.; Brandstetter, S.; Grolimund, D. Following dislocation patterning during fatigue. *Acta Mater.* **2016**, *112*, 184–193. [CrossRef]
12. Armstrong, R.W. Size effects on material yield strength/deformation/fracturing properties. *J. Mater. Res.* **2019**, 1–16. [CrossRef]
13. Zaiser, M.; Bay, K.; Hähner, P. Fractal analysis of deformation-induced dislocation patterns. *Acta Mater.* **1999**, *47*, 2463–2476. [CrossRef]
14. Hornbogen, E. Fractals in microstructure of metals. *Int. Mater. Rev.* **1989**, *34*, 277–296. [CrossRef]
15. Groma, I.; Bakó, B. Dislocation patterning: From micro-to mesoscale description. *Phys. Rev. Lett.* **2000**, *84*, 1487. [CrossRef] [PubMed]
16. Zhao, X.; Wu, J.; Chiu, Y.; Jones, I.; Gu, R.; Ngan, A. Critical dimension for the dislocation structure in deformed copper micropillars. *Scr. Mater.* **2019**, *163*, 137–141. [CrossRef]
17. Mandelbrot, B. How long is the coast of Britain? Statistical self-similarity and fractional dimension. *Science* **1967**, *156*, 636–638. [CrossRef] [PubMed]
18. Bak, P. *How Nature Works: The Science of Self-Organized Criticality*; Springer: Berlin/Heidelberg, Germany, 2013.
19. Iliopoulos, A. Complex systems: Phenomenology, modeling, analysis. *Int. J. Appl. Exp. Math.* **2016**, *1*, 105. [CrossRef] [PubMed]

20. Vinogradov, A.; Yasnikov, I.; Estrin, Y. Evolution of fractal structures in dislocation ensembles during plastic deformation. *Phys. Rev. Lett.* **2012**, *108*, 205504. [CrossRef]
21. Hähner, P.; Bay, K.; Zaiser, M. Fractal dislocation patterning during plastic deformation. *Phys. Rev. Lett.* **1998**, *81*, 2470. [CrossRef]
22. Hähner, P.; Zaiser, M. Dislocation dynamics and work hardening of fractal dislocation cell structures. *Mater. Sci. Eng. A* **1999**, *272*, 443–454. [CrossRef]
23. Sevillano, J.G.; Arizcorreta, I.O.; Kubin, L. Intrinsic size effects in plasticity by dislocation glide. *Mater. Sci. Eng. A* **2001**, *309*, 393–405. [CrossRef]
24. Koslowski, M.; LeSar, R.; Thomson, R. Dislocation structures and the deformation of materials. *Phys. Rev. Lett.* **2004**, *93*, 265503. [CrossRef]
25. Chen, Y.S.; Choi, W.; Papanikolaou, S.; Sethna, J.P. Bending crystals: Emergence of fractal dislocation structures. *Phys. Rev. Lett.* **2010**, *105*, 105501. [CrossRef]
26. Bakó, B.; Hoffelner, W. Cellular dislocation patterning during plastic deformation. *Phys. Rev. B* **2007**, *76*, 214108. [CrossRef]
27. Ludwig, W.; Cloetens, P.; Härtwig, J.; Baruchel, J.; Hamelin, B.; Bastie, P. Three-dimensional imaging of crystal defects bytopo-tomography'. *J. Appl. Crystall.* **2001**, *34*, 602–607. [CrossRef]
28. Tanaka, M.; Honda, M.; Mitsuhara, M.; Hata, S.; Kaneko, K.; Higashida, K. Three-dimensional observation of dislocations by electron tomography in a silicon crystal. *Mater. Trans.* **2008**, *49*, 1953–1956. [CrossRef]
29. Liu, G.; Robertson, I. Three-dimensional visualization of dislocation-precipitate interactions in a Al–4Mg–0.3 Sc alloy using weak-beam dark-field electron tomography. *J. Mater. Res.* **2011**, *26*, 514–522. [CrossRef]
30. Chen, C.C.; Zhu, C.; White, E.R.; Chiu, C.Y.; Scott, M.; Regan, B.; Marks, L.D.; Huang, Y.; Miao, J. Three-dimensional imaging of dislocations in a nanoparticle at atomic resolution. *Nature* **2013**, *496*, 74. [CrossRef]
31. Yamasaki, S.; Mitsuhara, M.; Ikeda, K.; Hata, S.; Nakashima, H. 3D visualization of dislocation arrangement using scanning electron microscope serial sectioning method. *Scr. Mater.* **2015**, *101*, 80–83. [CrossRef]
32. Madec, R.; Devincre, B.; Kubin, L. Simulation of dislocation patterns in multislip. *Scr. Mater.* **2002**, *47*, 689–695. [CrossRef]
33. Cui, Y.; Po, G.; Ghoniem, N. Size-Tuned Plastic Flow Localization in Irradiated Materials at the Submicron Scale. *Phys. Rev. Lett.* **2018**, *120*, 215501. [CrossRef]
34. Arsenlis, A.; Cai, W.; Tang, M.; Rhee, M.; Oppelstrup, T.; Hommes, G.; Pierce, T.G.; Bulatov, V.V. Enabling strain hardening simulations with dislocation dynamics. *Model. Simul. Mater. Sci. Eng.* **2007**, *15*, 553. [CrossRef]
35. Po, G.; Ghoniem, N. Mechanics of Defect Evolution Library, MODEL. Available online: https://bitbucket.org/model/model/wiki/home (accessed on 29 March 2019).
36. Ghoniem, N.M.; Tong, S.H.; Sun, L. Parametric dislocation dynamics: A thermodynamics-based approach to investigations of mesoscopic plastic deformation. *Phys. Rev. B* **2000**, *61*, 913. [CrossRef]
37. Po, G.; Mohamed, M.S.; Crosby, T.; Erel, C.; El-Azab, A.; Ghoniem, N. Recent Progress in Discrete Dislocation Dynamics and Its Applications to Micro Plasticity. *JOM* **2014**, *66*, 2108–2120. [CrossRef]
38. Cui, Y.; Po, G.; Ghoniem, N.M. A coupled dislocation dynamics-continuum barrier field model with application to irradiated materials. *Int. J. Plast.* **2018**, *104*, 54–67. [CrossRef]
39. Cui, Y.; Po, G.; Ghoniem, N. Suppression of Localized Plastic Flow in Irradiated Materials. *Scr. Mater.* **2018**, *154*, 34–39. [CrossRef]
40. Po, G.; Cui, Y.; Rivera, D.; Cereceda, D.; Swinburne, T.D.; Marian, J.; Ghoniem, N. A phenomenological dislocation mobility law for bcc metals. *Acta Mater.* **2016**, *119*, 123–135. [CrossRef]
41. Uchic, M.D.; Dimiduk, D.M.; Florando, J.N.; Nix, W.D. Sample dimensions influence strength and crystal plasticity. *Science* **2004**, *305*, 986–989. [CrossRef]
42. Schurch, R.; González, C.; Aguirre, P.; Zuniga, M.; Rowland, S.M.; Iddrissu, I. Calculating the fractal dimension from 3D images of electrical trees. In Proceedings of the 2017 International Symposium on High Voltage Engineering, Buenos Aires, Argentina, 28 August–1 September 2017.

© 2019 by the authors. Licensee MDPI, Basel, Switzerland. This article is an open access article distributed under the terms and conditions of the Creative Commons Attribution (CC BY) license (http://creativecommons.org/licenses/by/4.0/).

Article
Size Effects of High Strength Steel Wires

Kanji Ono

Department of Materials Science and Engineering, University of California, Los Angeles (UCLA), Los Angeles, CA 90095, USA; ono@ucla.edu; Tel.: +1-310-825-5534

Received: 6 February 2019; Accepted: 14 February 2019; Published: 17 February 2019

Abstract: This study examines the effects of size on the strength of materials, especially on high strength pearlitic steel wires. These wires play a central role in many long span suspension bridges and their design, construction, and maintenance are important for global public safety. In particular, two relationships have been considered to represent strength variation with respect to length parameters: (i) the strength versus inverse square-root and (ii) inverse length equations. In this study, existing data for the strength of high strength pearlitic steel wires is evaluated for the coefficient of determination (R^2 values). It is concluded that the data fits into two equations equally well. Thus, the choice between two groups of theories that predict respective relationships must rely on the merit of theoretical developments and assumptions made.

Keywords: Hall-Petch equation; Griffith equation; size effect; mechanical strength; pearlitic steels; suspension bridge cables

1. Introduction

Long span suspension bridges such as the Brooklyn Bridge (486 m main span, 1883) and Akashi Kaikyo Bridge (1991 m main span, 1998) owe their existence to high strength steel wires. The 4.7-mm diameter wires for the Brooklyn Bridge's main cables attained the tensile strength of 1.1 GPa in 1883 [1]. Over a century later, 5-mm diameter Akashi wires reached 1.8 GPa in 1998. The strength level increased to 1.9 GPa for 7-mm diameter steel wires used for the Hong Kong-Zhuhai-Macau Bridge of cable-stayed type, which was completed in 2018, while 2 GPa cable wires of 5- or 7-mm diameter have been available since 2015 [2]. Composition-wise, these wires are eutectoid carbon steels (with 0.8 to 1 wt % C) and are heat-treated to produce fine pearlitic microstructures during the isothermal phase transformation, known as patenting. The wires are next deformed during a series of cold drawing operations, resulting in wires of high strength with moderate ductility. Wires for various applications can be drawn down to smaller diameters, producing even higher strength. For example, ASTM A228 specifies music spring wires up to 3.3 GPa level, as the wire diameter decreases to 0.100-mm. In laboratory, the maximum strength reached 6.9 GPa for 1% C steel [3].

A recent article tracked the history of iron and steel usage for bridge construction [4]. Before the era of these huge suspension bridges with high strength steel wires as main cables, engineers had to use lower strength wrought iron wires for early suspension bridges. Examples from the US include the Wheeling Bridge (finished in 1849 and rebuilt in the 1860s) and the Niagara Falls Railroad Bridge (1855). Another choice for suspension members is the use of wrought iron chains. Truss and arch bridges also used wrought iron and steel (e.g., Eads Bridge, 1874). The use of cast iron was limited due to its structural deficiencies, but Iron Bridge at Coalbrookdale, UK (1781) remains a symbol of the Industrial Revolution. Still, most of these bridges were built in the 19th century or later. In order to locate the traces of pre-Industrial Revolution iron bridge building, researchers were required to go to China and South/Central Asia, where 2000 years ago ferrous metallurgy was more developed compared to the rest of the ancient world. Historical records exist in the form of travelogues written by Chinese Buddhist monks, including Faxian and Xuanzang. They trekked from China to India in the 4th to 7th century, respectively, and had to travel through the Pamir and Hindu Kush mountains, where they recorded their travel going over iron bridges or iron chain bridges. Western bridge engineers [1,5] often cited a Chinese iron bridge built in 56 or 65 AD (based on a 17th century history book by Kircher), but this has no support from Chinese bridge historians and no historical record exists [6]. Another attribution was an iron bridge built in the year 206 BC as a part of war efforts for the succession of the Qin Dynasty. However, the source, Sima's historical volumes, only mentions that bridges were built. Besides, no archaeological evidence has been uncovered. An iron bridge called Ji-Hong, built in 1475, was the most credible early example, but was destroyed in a 1986 landslide [7]. See details on the history of iron bridges in [4].

Modern suspension bridges built in the US and Europe were possibly inspired by Pope's 1811 book [8], which described a suspension bridge in Bhutan with a detailed illustration [4]. Several 15th century iron bridges existed in Bhutan until the 1960s, though only a reconstructed bridge is left today. Pope was also remarkable for his technical foresight; he warned of the lack of redundancy and of instability against vibration of suspension bridges, long before these became serious issues. In 1816, a simple pedestrian bridge was built in Philadelphia, using two three-strand twisted cables with wooden planks; it did not hold during its first winter storm due to inadequate design [5]. More durable suspension bridges were built in the early 1820s in Francophone Europe, one of which still exists today [9]. These had parallel wire design for the main cables, which ensured a high loading capacity. This technology spread in surrounding regions for the next 30 years and many similar bridges were built. However, its weakness against wind-driven oscillations manifested as a major disaster at Angers Bridge in 1850, which killed 226. A similar fate fell on the Wheeling Bridge in 1854 as it was built by Ellet using the French technology [5]. In the US, stiffening of bridge structures allowed continual development of suspension bridges, which were particularly valued in the rapidly developing western states.

These bridge constructions and other industrial activities spawned breakthroughs in wire-making in mid-19th century England [4]. Strong iron wires for musical instruments were first made in Augsburg, Germany in 1351 and German firms dominated the industry until 1834 [10]. Webster of Birmingham, UK used Mn-containing steel and doubled the wire strength in 1825. In the late 1840s, Horsfall, also from Birmingham, introduced isothermal phase transformation process, now known as patenting, and raised the wire strength further. The name "patenting" originated from the fact that the new process received British patents (at least three are on record) in the 1850s and Webster and Horsfall (merged in 1855) marketed their wires as "patent steel wires". They found many industrial applications for the high strength steel wires at 1 GPa level, including 1860 trans-Atlantic telegraph cables that required 1600 t per installation. This period also was the time of innovation in steelmaking with Bessemer and open-hearth processes. By the 1880s, steel strength attained 1.4 GPa for 4.7-mm diameter wires suitable for bridge cable uses [11].

One overlooked breakthrough in Horsfall's invention is his choice of starting stocks for his wire drawing. Conventional wisdom is to use more ductile annealed wire rods, but he chose to use patented stocks of higher strength. Because of thinner cementite (iron carbide) layers in patented wires, higher drawing strain can be imposed, producing stronger final wire products. It was remarkable that Horsfall developed his process without the microstructural knowledge developed many years later; he deserves our appreciation for this contribution as well.

The aforementioned historical background leads to the main subject of discussion, namely, what makes the drawn wires strong. For more than 50 years, it has been clear that smaller pearlite spacings results in higher strength [12]. Yet, discussion continues as to the origin either from theories underlying the Hall-Petch equation for the tensile strength, $\sigma_{ts}(e)$,

$$\sigma_{ts}(e) = \sigma_o + k/\sqrt{d}, \tag{1}$$

or from those supporting the Griffith equation:

$$\sigma_{ts}(e) = A + B/d. \tag{2}$$

Here, σ_o, k, A, and B are constants and d is a length parameter. For the Hall-Petch equation, d is the grain size, while for the Griffith equation d is diameter of wire or fiber. Many studies and reviews examined experimental observations and favor one or the other. However, most past data evaluation lacked statistical aspects. The aim of this work is to provide data comparisons with statistical parameters. Results indicate that currently available experimental data is inadequate to decide one or the other equation to be the only valid relationship. Thus, a final decision rests on the robustness of the theories that support a correlation.

2. Survey of General Size Effects

Strength increases with diametrical reduction of drawn wires were controlled by intermediate (or interpass) annealing, even though the beginning of this procedure is obscure. Wright [13] suggested that, by 5th century BC, Persians used iron draw plates and interpass annealing to make 0.55-mm bronze wires. The size effects of iron wire strength were first recorded in 1824 by early suspension bridge builders in France and Switzerland [8]. The Seguin brothers conducted 80 tests in France, whereas Dufour conducted 22 in Switzerland using wires obtained locally. Wire diameter ranged from 0.59 to 5.94 mm. Dufour fitted his results of 22 tests to an inverse diameter relation in terms of the tensile strength σ_{ts} and diameter d (in mm) using Equation (2) with A = 411 MPa and B = 276 MPa-mm (R^2 = 0.887). The strength results for diameters less than 2.8 mm were plotted in Figure 1a by red points. (Larger diameter data was inconsistent and omitted.)

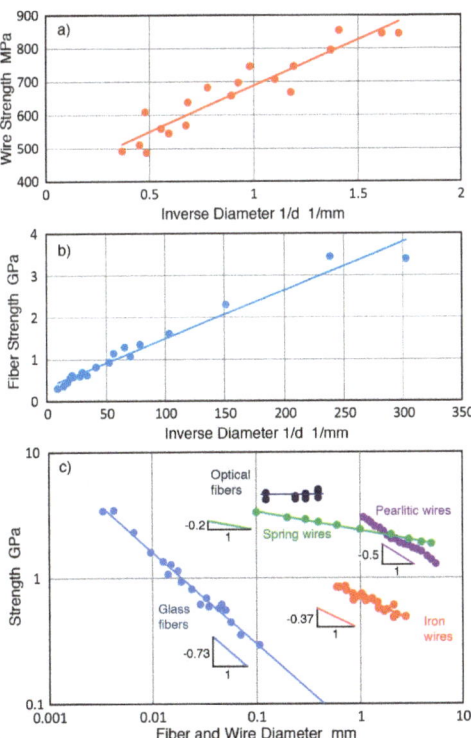

Figure 1. Strength versus diameter of fibers or wires. (**a**) Dufour and Seguin brothers' data (1824) from [9]. (**b**) Griffith glass fiber data (1921) from [14]. (**c**) Log-log plots of five data sets as indicated data from [4,9,14–16]. See text.

In his monumental paper on fracture criterion, Griffith [14] cited Karmarsh [17], who in 1858, obtained the same inverse diameter relation for metal strength. Griffith used the inverse diameter relation or the Griffith equation for describing his results on the strength of freshly drawn glass fibers, which are also plotted in Figure 1b by blue points. The strength levels reached much higher levels and d values much smaller; the data fits to Equation (2) with A = 329 MPa and B = 11.56 MPa-mm (R^2 = 0.967). Iron wire and glass fiber data sets also fit a power law relation of

$$\sigma_{ts}(e) = C\, d^{-n}, \tag{3}$$

with a constant, C, and an exponent, n. The 1824 iron wire data yields n = 0.372 (R^2 = 0.892) and Griffith glass fiber data gives n = 0.73 (R^2 = 0.982). These are in red and blue points in Figure 1c. Closeness of respective R^2 values indicates each data set fits to either the Griffith equation or power law. The strength values of ASTM A228 specification are plotted in Figure 1c as green points, giving a power-law fit with n = 0.15 (R^2 = 0.994). However, these points fail to follow the Griffith equation. This plot represents five other data sets specified in ASTM A313 for 304, 316, 17-7PH (as-drawn or with aging), and XM28 steels. These have slightly lower exponents of 0.09 to 0.12. The spring wires are to be fabricated with additional deformation and the specified values are lower than their respective upper limits. While not analyzed for power-law fit, several other ASTM standards cover alloy steels and bronze wires in A229, A232, A877, and B159. Similar fits as in A228 and A313 are anticipated.

Another data set [15] for pearlitic steel wires (purple points in Figure 1c) fits to Equations (2) and (3) with R^2 levels of 0.99 for the Griffith equation and 0.987 for power law with n = 0.485. This Ochiai data set (for steel F) is prototypical for all other pearlitic steel wires, as will be shown below [15]. In this case, each point represents an as-drawn condition from a single starting diameter and a smaller diameter resulted from a higher drawing strain. Another data set was given in Ochiai (steel G) [15] and has n = 0.195, comparable to the A228 data. The wires in this group were drawn with interpass patenting to attain high drawing strain of up to 6.4.

Beyond the above cases (and those in Table 1), only a few studies dealt with the size effect on metal wires. Rubenstein [18] examined size effects on Ni wires with different microstructures. However, data scatters are large and it is difficult to draw definitive conclusions. Riesch et al. [19] studied size effects on W wires and collected previous results from the literature. While the authors contend the results fit the Hall-Petch relation, the data set also fits with the Griffith equation with comparable R^2 values of 0.829 and 0.793, which may be called moderate fits at best. Metal conditions varied and the data came from eight different articles. Thus, the low R^2 values are expected and the tungsten results hardly contribute to our discussion. Actually, two more sources of 19th century music wires exist, but are omitted here as the ranges of diameters were limited [10,20].

Griffith's glass fiber data has been explained in terms of existing flaws with Weibull statistics. In this approach, the probability of failure P is given by

$$1 - P = \exp\{-(\sigma/\sigma_o)^m \cdot d^h \cdot L\}, \tag{4}$$

where m is the shape parameter (also called Weibull modulus), σ_o scale parameter, h diameter dependency parameter, d diameter, and L sample length, respectively [21]. In the original Weibull theory [22], h = 1 for surface flaw-controlled failure and h = 2 for volume-controlled failure. From Equation (4), the average fiber strength $<\sigma>$ is given by

$$<\sigma> = K \cdot L^{-m} \cdot d^{-h/m}, \tag{5}$$

where K = $\sigma_o \cdot \Gamma(1 + 1/m)$ and $\Gamma(x)$ is the gamma function. While $<\sigma>$ value is easily deduced in experiment, the importance of this equation comes from the correlation between the power-law exponent n and the Weibull shape parameter. When both n and m are measured, h value is determined. In order to clarify the physical meaning of h, it is desirable to collect more experimental data beyond

Zhu's study [21] though this task is challenging. As of now, the meaning of h is unclear when it is not 1 or 2.

For today's common glass fibers for composite reinforcement, m is either 3 or 4. For the Griffith fibers, surface flaws are assumed to be the fracture origins ($h = 1$) and the expected value of $m = 1/n = 1/0.76 = 1.3$. This m value appears reasonable for hand-drawn fibers in the 1920s. When careful process control is practiced, the size effects of glass fiber are absent, as demonstrated by Otto [23]. Current optical glass fibers that are well protected by polymer coating also exhibit no size effect of their strength [16]; 18 samples of 0.125- to 0.4-mm diameter ($L = 1$ to 2.5 m) showed the average strength of 4.58 ± 0.29 GPa. These points are plotted in Figure 1c with dark blue points.

Zhu et al. [21] tested seven types of ceramic fibers, measuring m and h parameters. Their fibers included alumina, sapphire, Si_3N_4, SiC, Nicalon, and Nextel fibers. The values of m ranged from 2.5 to 14 and those of h ranged from 1.4 to 19. Obviously, h values are not limited to 1 or 2 of the original Weibull theory. However, in the two cases where h was 13 and 19 also showed a large m, giving an h/m value of ~1. All the h/m values were between 0.5 and 1.3. Further work is needed if this is significant. A recent study [24] on the synthesis of SiCN fibers with electron irradiation included Weibull parameter determination as well as size effects. Fiber diameters ranged from 28 to 95 µm and over 50 samples were used with each starting material. The value of m was 4.46, while a power-law fit yielded $n = 0.57$ (they fitted the data with an exponential function, but large scatter in the data allows fitting with either). From the m and n values, $h = 2.54$ resulted. This is close to volume-controlled flaw effect of the Weibull theory. Since their fiber fabrication processes are always inside a vacuum chamber, surface flaws may be minimized. This is consistent with $h = 2$. Other findings on the h parameter will be discussed in Section 4.2.

Weibull analysis is infrequently conducted for metallic materials since m values have been expected to be around 100, although definitive studies seem to be absent. An approximate method for Weibull modulus estimation discussed in Appendix A found two sets of bridge cable wires, before service, having m values of 110 and 124. A recent study obtained an m value of 56 to 72 for stainless steels [25]. Bridge engineering guidelines [26] noted m of 70 for slightly corroded steel cable wires. Cable wires, new and lightly damaged, have m values above 50, but more severely corroded wires had reduced m, going down to m of 10 to 30. For the nearly 100-years old Williamsburg Bridge cable wires, m was found to be 16.0 [4], while a still older data set from 1886 showed an m of 13.7 [11]. Thus, old or corroded wires have low m values, while undamaged high strength steel wires possess high m values and size effects predicted by the Weibull theory are minimal. In the metal wire cases, m is large and the power-law exponent corresponds to h/m. Although h values have not been determined for metals, it is prudent to use Weibull's unity value for surface flaw critical cases [22]. Then, the observed power-law exponents in Figure 1c are primarily contributed by plastic deformation, not by the Weibull size effects.

3. Size Effects of Pearlitic Steels

The extremely high strength levels of patented and cold drawn eutectoid and hypereutectoid steel wires have been studied for many years and resulted in numerous patent filings. The quest for the clarification of their origins accelerated with the availability of transmission electron microscopy (TEM) in the 1960s, atom probe microscopy (APM) in the 1980s, and 3D-APM since the 2000s [27–29]. Embury and Fisher [30] were the first to use TEM for correlating pearlite lamellar spacings to the strength and drawing strain. They established the basic understanding of microstructural effects, as well as the correlation between the lamellar spacings and wire diameter. For the correlation, they chose to use the Hall-Petch relationship or Equation (1), but without excluding other possibilities. Langford [31,32] provided more detailed examination of pearlite strengthening. Langford and Cohen [33] examined deformed iron and they chose Equation (2) or the Griffith equation. They related the strength and the inverse of dislocation cell size on the basis of the Frank-Read source operation. Note that d in Equation (2) can be directly replaced with pearlite spacing (in the diametral directions only), as their equivalence was established earlier [30]. Marder and Bramfitt [34] used the Griffith equation approach in describing the strengthening effects of thermally varied pearlite spacings. Some subsequent studies followed Embury-Fisher's choice of the Hall-Petch equation, most recently by Borchers and Kirchheim [35], while others favored the Griffith equation [36,37]. One persistent finding is that the Hall-Petch equation produces negative or low strength values when $1/\sqrt{d}$ term decreases [34]. The differences between the two interpretive approaches are the underlying theories of strength determination in microscopic lamellar structures. Unlike the grain sizes in the tens of μm, however, TEM cannot provide clear-cut evidence in support of one theory from another in heavily deformed pearlite. Besides, high dislocation densities always make it difficult to resolve critical events. These uncertainties may be clarified if experimental size effects of strength can show one approach giving a better fit. Surprisingly, all studies examined here did not conduct a direct comparison of data fitting to the two equations. This is the main goal of the present evaluation of published strength dependences on drawing strains, which correlate to the pearlite spacings.

In this part, 19 publications and one unpublished doctoral dissertation were examined [3,15,30–32,36–50]. From their graphical data, values of tensile strength σ_{ts} and of drawing strain e (or diameter d) were obtained. As such, 18 data sets of σ_{ts} versus e are plotted in Figure 2. Several of them were already given in tabular form, but most were estimated from figures. In some works, multiple results were presented and two or three representative results were used. When the number of data points were less than eight, these data sets were analyzed, but not plotted in Figure 2 or used in calculating averages. These plots demonstrate consistency of observed hardening behavior, though deviations become large when e values exceed 4. Most had the starting strength of 1.3–1.5 GPa, but two curves had low starting strength (1–1.2 GPa) and one had a higher value (1.7 GPa). These plots show that all the data sets behave as expected for high C steels. Table 1 presents the articles evaluated in chronological order, represented by the first author and year of publication. Next column lists the sample counts. The fourth column gives the exponent n, obtained by plotting the strength against diameter. When the starting diameter is unknown, 5 mm was used. The next three columns provide R^2 values obtained for n and by plotting the strength against $\exp(e/2) = d_o/d$ or $\exp(e/4) = (d_o/d)^{0.5}$, where e represents the true strain. The last two gives notes and reference number. The data for studies with small sample counts are separated to the bottom as the data significance is lower and n values are omitted. In statistical terms, even twenty samples are inadequate sample counts, but technically this is the typical upper limit in wire drawing facilities.

Table 1. Statistical data for comparison among the three types of fitting.

Authors	Year	Sample Count	n for TS Versus d^{-n}	R^2 for n	R^2 for TS-exp(e/2)	R^2 for TS-exp(e/4)	Notes	Ref.
Embury	1966	12	0.551	0.975	0.969	0.981	-	[30]
Langford	1970	17	0.469	0.995	0.971	0.995	-	[31]
Langford	1970	13	0.482	0.990	0.990	0.990	w/o e > 4	[31]
Yamakoshi	1973	19	0.507	0.994	0.998	0.993	steel B	[47]
Yamakoshi	1973	15	0.514	0.995	0.997	0.993	steel C	[47]
Yamakoshi	1973	15	0.507	0.993	0.997	0.992	steel F	[47]
Langford	1977	9	0.505	0.996	0.986	0.997	strip	[32]
Kanetsuki	1991	10	0.397	0.980	0.991	0.982	-	[37]
Ochiai	1993	18	0.485	0.987	0.990	0.987	steel F	[16]
Nam	1995	13	0.504	0.978	0.978	0.986	-	[41]
Choi	1996	10	0.562	0.972	0.965	0.968	-	[38]
Makii	1997	22	0.459	0.989	0.994	0.992	bridge cable	[48]
Makii	1997	15	0.396	0.976	0.973	0.974	tire cord	[48]
Tashiro	1999	14	0.471	0.993	0.975	0.991	0.5 mm	[36]
Tashiro	1999	10	0.480	0.988	0.996	0.985	w/o e > 5	[36]
Tashiro	1999	12	0.468	0.987	0.996	0.985	1.0 mm	[36]
Buono	2002	9	0.488	0.994	0.991	0.994	-	[46]
Zelin	2002	18	0.538	0.993	0.998	0.991	-	[42]
Tarui	2010	16	0.507	0.992	0.964	0.990	-	[49]
Tarui	2010	13	0.509	0.985	0.996	0.981	w/o e > 4	[49]
Li	2014	8	0.447	0.974	0.844	0.945	-	[3]
Average of n and R^2	-	-	0.488	0.987	0.979	0.985	-	-

Table 1. Cont.

Authors	Year	Sample Count	n for TS Versus d^{-n}	R^2 for n	R^2 for TS-exp(e/2)	R^2 for TS-exp(e/4)	Notes	Ref.
Std deviation	-	-	0.042	0.008	0.033	0.012	-	-
Pepe	1973	5	Linear fit	-	0.936	0.957	-	[45]
Kim	1992	7	0.544	0.990	0.991	0.982	-	[43]
Maruyama	2002	5	0.508	0.991	0.987	0.996	w/o e > 4.5	[39]
Goto	2007	3	0.439	0.995	0.984	0.996	-	[44]
Zhang	2011	4	0.526	0.990	0.998	0.992	-	[40]
Zhang	2016	4	0.755	0.983	0.985	0.988	PS used	[50]
Li	2014	5	0.538	0.995	0.990	0.996	w/o e > 5	[3]

Author: The first author only is shown; w/o e > 4: without data for e larger than 4; 0.5 (1) mm: Data for 0.5 (1) mm diameter wires; PS: pearlite spacing; Ref.: reference number.

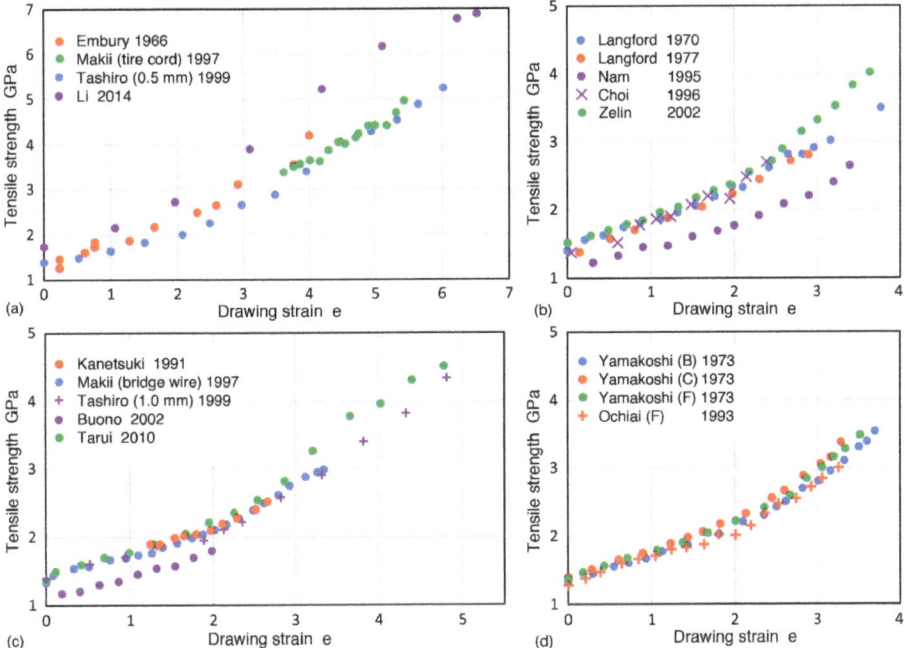

Figure 2. Tensile strength, σ_{ts} versus drawing strain, e, for 18 data sets. (**a**) Data from [3,30,36,48]. (**b**) Data from [31,32,38,41,42]. (**c**) Data from [36,37,46,48,49]. (**d**) Data from [15,47]. See inserts for symbols used.

Figure 3 plotted the data for Makii et al. [48] for the tensile strength of a pearlitic eutectoid steel (0.8% C) against diameter d (in green), normalized inverse diameter d_o/d (in blue) and $\sqrt{(d_o/d)}$ (in red). The $\sqrt{(d_o/d)}$ scale is doubled for a better comparison. This data set is shown first since its sample count is 22, the largest. These plots are accompanied by a power-law (green) curve for σ_{ts} versus d with exponent $n = 0.507$ and linear fits for the other two (in blue and red). That is, red points are represented by Equation (1) and blue points by Equation (2). While some deviations are observed, all three length parameters provide excellent fits with R^2 values (or coefficients of determination) of 0.992 (d), 0.994 (d_o/d) and 0.992 ($\sqrt{(d_o/d)}$), respectively. Thus, either Equations (1) or (2) can represent the observed size dependence equally well. The power law fit represents the strength increase due to cold drawing, their exponents and n values are also tabulated in Table 1 for all other data sets. The n values ranged from 0.4 to 0.55, and their average was 0.488 with R^2 value of 0.987. Therefore, the d dependence (with $R^2 = 0.985$) is essentially the same as the Hall-Petch equation.

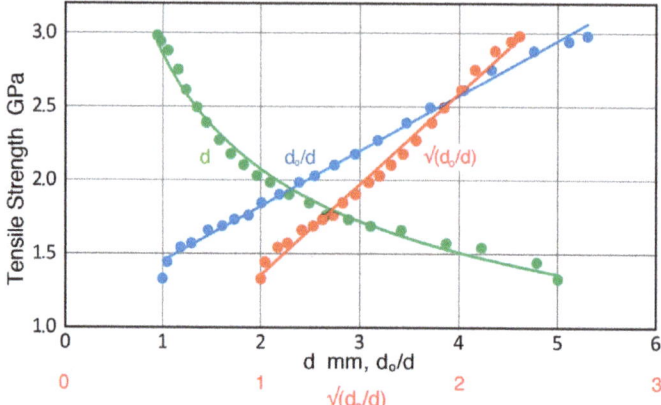

Figure 3. Tensile strength versus length parameters. Makii data [48]. Green plot: σ_{ts} versus d; Red: σ_{ts} versus $\sqrt{(d_o/d)}$, Blue: σ_{ts} versus (d_o/d).

The remaining data sets also exhibited a similarly good match of R^2 values, as shown in Table 1. Plots with the same format as Figure 3 are given in Figure 4 for eight more data sets. These show similar features observed in Figure 3. The average R^2 values were 0.979 and 0.985 and the difference of 0.006 was much less than the standard deviation. Collectively and individually, no differentiation can be made between Equations (1) and (2). In some cases, fits improved further when high strain data ($e > 4$) were omitted. The strains above 4 produced less work hardening and this has been attributed to cementite dissolution and other causes. Li et al. [3] observed a transition from lamellar structure to nanosized subgrain structure at $e = 3.8$. This high strain effect was most pronounced in Li data, as it included strains up to 6.5. When the highest three points are removed, R^2 values become comparable to other data sets (0.990 and 0.996), as the bottom line on Table 1 shows. The fits are shown on Figure 5. Note that the $\sqrt{(d_o/d)}$ scale is expanded five-fold.

From the comparison of available size effect data on pearlitic steel strength, it appears difficult to distinguish fits to Equation (1) or (2). At the same time, the power-law fit is just as good as the Hall-Petch or Griffith equations. While we have used a power-law expression for modeling stress-strain relations for a long time, this has not been applied in connection to length parameters. This will be examined next if a new way to evaluate the observed data can be found.

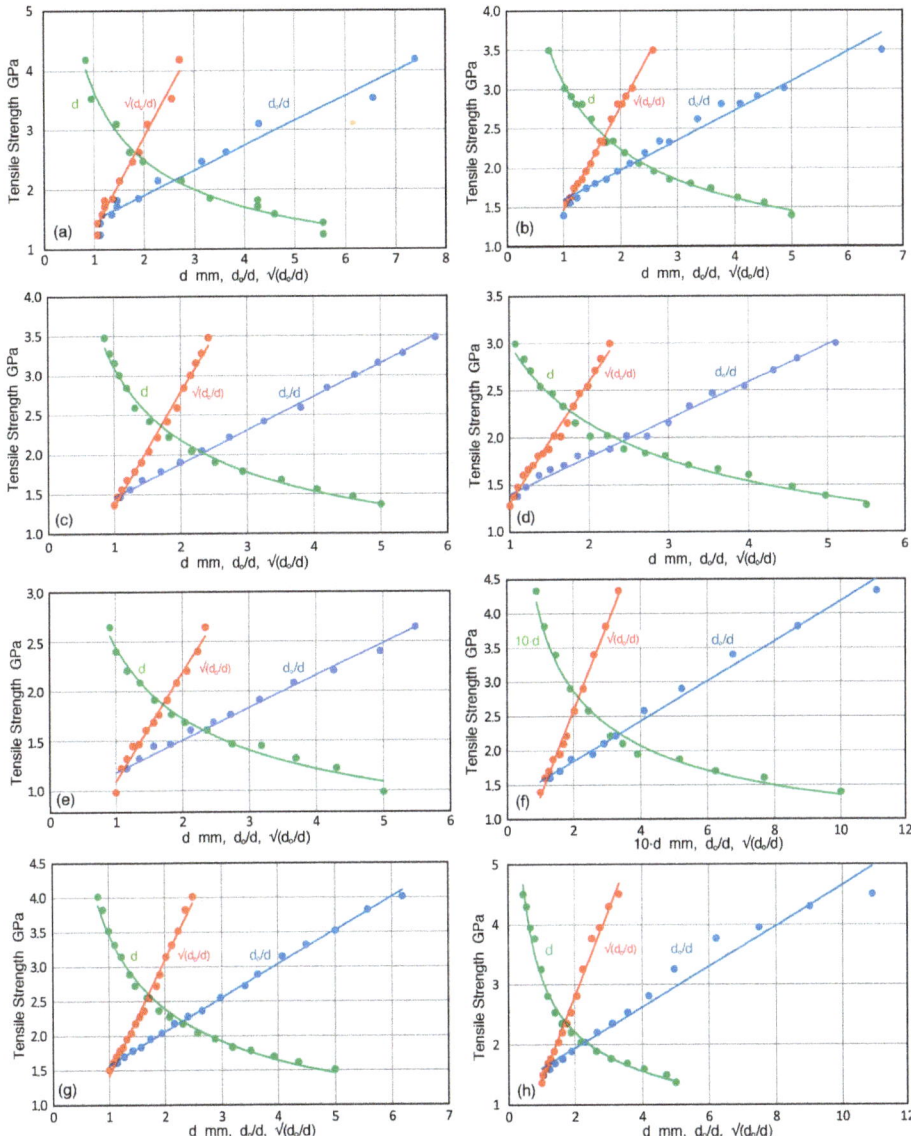

Figure 4. Tensile strength versus length parameters. (**a**) Embury data [30], (**b**) Langford [32], (**c**) Yamakoshi [47], (**d**) Ochiai [15], (**e**) Nam [41], (**f**) Tashiro [36], (**g**) Zelin [42], and (**h**) Tarui [49]. Green plot: σ_{ts} versus d; Red: σ_{ts} versus $\sqrt{(d_o/d)}$, Blue: σ_{ts} versus (d_o/d).

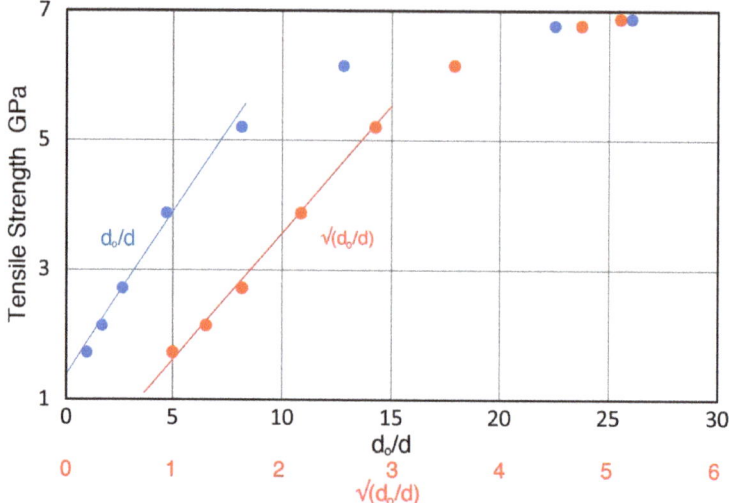

Figure 5. Tensile strength versus length parameters. Fits improve substantially without high strain data points. Li data [3].

4. Discussion

4.1. Diameter Dependence

When Equation (3) is rewritten with $n = 0.5$ as

$$\sigma_{ts}(e) = Cd^{-0.5} = C'd_o^{0.5}d^{-0.5} = C'(d_o/d)^{0.5},$$

we get

$$\sigma_{ts}(e) = C'[\exp(e/2)]^{0.5} = C' \cdot \exp(e/4). \tag{6}$$

That is, we arrive at Equation (1) since $d \leq d_o$ or $\exp(e/2) \geq 1$ and $\sigma_{ts}(1) = C'$. Similarly, for $n = 1$, we can rewrite the power law using $C = C''/d_o$ as

$$\sigma_{ts}(e) = C'' \cdot \exp(e/2). \tag{7}$$

This is a form of Equation (2). Note that both equations are not defined for d_o/d less than 1 and C is the starting strength for the wire of diameter d_o. Equations (6) and (7) are plotted in Figure 6 with blue and red curves. Considering these two expressions, when observed data fits to a power law with the exponent of 0.5 as seen in the previous section, the Hall-Petch relation appears to be the proper equation. This is because the strength-diameter plots in Figures 3 and 4 cannot be fitted to the inverse diameter function or Equation (7). However, both Hall-Petch and Griffith equations have a constant term, σ_o and A. An example of such a Griffith equation, or

$$\sigma_{ts} = 0.75 + 0.3/d, \tag{8}$$

is also plotted (in green points) in Figure 6. Here, the units are in GPa and mm. Equations (6) and (8) are close and R^2 value was 0.978. Even though this equation has not been optimized for a better fit, this R^2 is almost identical to the average R^2 value for $1/d$ plots in Table 1. Therefore, it is necessary to conclude that experimental size effects on strength can be represented by two different functional forms. In fact, other functional forms have not been excluded.

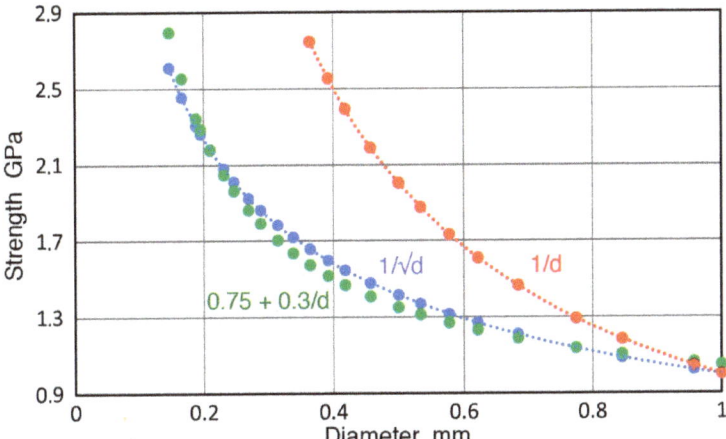

Figure 6. Strength versus diameter using three different representations.

4.2. Weibull Size Effects

The average strength of a data set that follows Weibull Equation (4) can be calculated by Equation (5). This depends on the sample length as $L^{-1/m}$ and the diameter as $d^{-h/m}$. For optical fibers, tests of up to 20-m length were used to predict fiber strength for $L = 100$ km, relying on the $L^{-1/m}$-dependence [51]. For glass fibers used in cables, $m < 10$, while laboratory m values are 50–70, reaching as high as 120. For carbon fibers made from polyacrylonitrile showing $m = 4$, $L^{-1/m}$-dependence was shown by Tagawa and Miyata [52]. In this case, diametral dependence was found to follow $d^{-1.18}$, giving $h = 4.7$ [52]. Separately, Tagawa found anisotropic size effects that can best be interpreted by giving a different m value in the radial direction, or $m_r = 0.45$. If this is assumed to be valid, h becomes 2.63, which is comparable to the case of SiCN fibers [24]. Another SiCN (Tyranno) fiber study [53] showed $L^{-1/4.7}$, while the Weibull modulus was 4.3, giving a good match. It is possible that a wide range of h values found in ceramic fibers [21] may come from different fabrication methods, compared to the melt processing for glass fibers or the precursor pyrolyzing processes of carbon (also Nicalon and Tyranno) fibers.

For organic fibers, such as polyethylene and Kevlar, more size effect studies were made, accompanied by Weibull analysis of strength [54–57]. Wagner [54] examined nine types of fibers for their diametral dependence, considering five functional behaviors (including Equations (1) and (2)). No preferred dependence emerged, however. For ultra-high strength polyethylene fibers, Schwartz et al. [55] found no length effect, while Smook et al. [56] fitted their data to $1/\sigma_{ts} = A' + B'\sqrt{d}$ relationship. Their data can be described by Equation (2) with A = 1.55 GPa and B = 53.7 GPa-μm ($R^2 = 0.985$). In another study [57], four similar ultra-high strength polyethylene fibers (including Spectra 900 and 1000) showed power-law fit (Equation (3)), but the exponents varied from 0.46 to 2.13. Most polymer fibers appear to suffer from geometrical nonuniformity, which affects statistical comparison. No consensus view has emerged on their size effect. It appears these fibers need to be treated separately from metallic wires.

Fracture problems related to corrosion and fatigue are as important as strength, but not covered in this study. Four recent works are listed as references [58–62]. For example, most fracture started from corrosion pits [60]. Unfortunately, a misconception introduced in the Silver Bridge disaster investigation [63], i.e., steel corrosion induced by hydrogen sulfide, is still invoked as a source of possible hydrogen embrittlement by bridge experts [62]. The hydrogen-sulfide hypothesis for bridge steel fracture was shown to be untenable [4]. Zinc (or iron) and acidic aqueous environment are adequate as hydrogen sources. Moreover, hydrogen embrittlement has been convincingly discounted for cable wires by detailed studies and the level of hydrogen in corroded steel wires were shown to be less than 0.2 ppm, which is insufficient for causing delayed fracture [64–66]. An interdisciplinary approach is essential.

4.3. Strength Partition

Since pearlite consists of ferrite and cementite lamella, its strength depends on both phases and how the two phases are distributed. Embury and Fisher [30] examined and discarded the rule of mixture approach as it was developed on the iso-strain assumption of fibrous mixtures. This was based on the ferrite and cementite strength as the base strength. Since then, more possible strengthening mechanisms have been proposed and reviewed [3,35,50]. Some critical issues, like the deformation and fracture of cementite, have been studied further. Fang et al. [67] showed with high resolution TEM dislocation induced shearing of cementite lamella at e below 1.5, while nano-size particulate rotation within the cementite lamella contributes to their thinning at higher strain. These observations give rational explanation of microstructural developments within deformed pearlite. Segregation of freed C atoms to dislocations is another source of potential strengthening. As ferrite is of bcc structure, Peierls barrier dictates dislocation mobility. The enhancement of the Peierls barrier by C interstitials and by carbide precipitation [68] was theoretically shown in 1970 [69], by unifying the Peierls and dispersed barrier strengthening mechanisms. Because of the complexity of multiple and interacting strengthening effects, the development of a unified theory for deformed pearlitic wires will take more time to develop. For analysis of what is known, an artificial intelligence approach may be of use. However, a recent study using neural networks [70] is a variation of pattern recognition analysis [71]. The black-box nature of neural networks is unsuited for developing new theoretical understanding of complex strengthening behavior.

4.4. Pearlite Spacings

Since the introduction of TEM in the materials research, earlier microscopic studies have often been overlooked. Of interest for the present discussion are two papers by Gensamer et al. [72,73]. They used optical microscopy (presumably with photographic enlargements) and determined pearlite spacings varied by using different isothermal phase transformation temperatures. The magnification reached 2000 to 6000 times and the smallest pearlite spacing was slightly below 0.1 µm. This spacing data set was corrected by a factor of 0.5 because of their use of the random intercept method [74]. Correlations with the mechanical parameters were obtained in terms of the logarithm of pearlite spacing [73]. Their tensile strength data (in red points) is plotted against the pearlite spacing, the inverse pearlite spacing, and the inverse square-root of pearlite spacing in Figure 7a–c. In addition, most of the available data in the literature was plotted, leaving out low resolution studies. Among these works, two studies varied transformation temperatures [34,75]. Others used cold drawing to reduce the pearlite spacings, starting from Embury and Fisher [30] (in dark blue squares). Their pearlite spacing data was read from the smallest spacings in TEM photographs since they only reported cell sizes. The cell size data was about twice the pearlite spacings and not used here. Further given are data from Langford [32] (in blue +), Tarui [49] (in green point), Zhang et al. [40,50] (in red and green X), Li et al. [3] (in green triangles), and Takahashi et al. [29] (in red triangle). Most studies [3,30,32,40,49,50,75], used TEM, while SEM [49], two-surface replica method [34], and atom probe microscope [3,29] were also used.

Figure 7a shows the tensile strength versus pearlite spacing from various studies using only the data from direct determination. Data scatter is larger than those from strength and drawing strain, reflecting experimental difficulties dealing with sub-micron dimensions. At 1 to 2 GPa strength levels, a spread of a factor of four is found in terms of spacings (as the strength data is expected to behave more consistently). Elsewhere, a factor of two to three is typical. Because of the proportionality between wire diameter and pearlite spacing and the observed power-law relation between the diameter and tensile strength, as observed in Section 3, a power law with an exponent of −0.5 (a straight line with slope of −0.5 in the log-log plot) is expected. This is drawn as the fitting line (in green), resulting in R^2 = 0.833. The data fitting by regression yielded the slope of 0.529, with identical R^2 = 0.833. Atom probe data [29], shown with a red triangle (plus several more points from [3] not shown) indicates consistency with TEM studies, but two points from Embury [30] were off by 2x or more. (These three points are shown, but were not included in R^2 calculations.) Figure 7b gives plots of the tensile strength versus inverse pearlite spacing. Identical symbols are used, as they are in Figure 7a, where most data points are for the pearlite spacings larger than 50 nm. The least-square fit with Equation (2) (shown by a blue line) resulted in R^2 = 0.853, indicating a slightly better fit than the power-law fit. Figure 7c illustrates the case using the variable of inverse square-root of pearlite spacing, again with identical symbols. The least-square fitting line is drawn (in red), producing the best fit among three plots with R^2 = 0.868. This is the fit to Equation (1) or Hall-Petch relation. This $1/\sqrt{}$(pearlite spacing)-fit is statistically close to the inverse pearlite spacing fit (equation (2)) in Figure 7b with a difference in R^2 of 0.015. As shown in Table 1, it was 0.01 in previous comparison. Because of data scatter, data fittings again allow one to support two groups of theories as found earlier.

This spacing data compilation relied on the published values. In most works, average pearlite spacings were reported, while some used the average of finest ferrite spacings [40,49]. In the case of Embury's data [30] used in Figure 7, the two finest ferrite spacing values were read from their TEM results. Ferrite spacings are around 90% of the corresponding pearlite spacing and this should be used for strength calculations [49,50]. Among the published studies, Langford [32] conducted most comprehensive TEM examination using 15 drawn wire samples. Recently, Zhang et al. [50] reported advanced TEM studies on ten drawn wire samples. Results of these two studies overlap well in Figure 7, especially below 60 nm. These data also agree well with the results of Marder and Bramfitt [34] above 60 nm, who used 25 samples from heat treatment. These three studies provided 2/3 of the data examined here. Note that the strength data for Zhang data used for Figure 7 was interpolated from four tensile strength values given in [50]. When only these three data sets are used, R^2 values increase to 0.874, 0.908 and 0.901, corresponding to the three plots in Figure 7. Thus, the difference in R^2 values between Equations (1) and (2) becomes 0.007 and supports the conclusion of data equivalency.

Vander Voort and Roosz [76] examined several measurement methods for pearlite spacings and concluded that the average values are consistent among four methods studied. They also noted that the finest spacing values were 40–50% of the average. This explains the factor-of-two difference observed in Figure 7 between high and low spacing groups. For the phase transformation studies, the global averages are appropriate, but for the strength-pearlite spacing correlation, the finest spacings control the maximum resistance to deformation. It is hoped that future pearlite studies consider this aspect in addition to increasing sample counts for statistically robust measurements. Because of ease in sample preparation, SEM methods are best for achieving adequate sampling. Tarui (private communication) noted that when 20–30 SEM fields of view are analyzed for finest spacings using Tashiro-Sato method [77], a convergence is obtained for a single finest ferrite spacing from about five fields. He used this value as the ferrite spacing, which satisfies statistical requirements. The smallest value reported using this method was 58 nm [49]. Since 100 nm range is routinely achievable, 5–10 nm measurements should be feasible. For instance, an old SEM image of as-patented 0.8C steel was kindly supplied by Toshimi Tarui of Nippon Steel Sumitomo Metals (NSSM). It was taken at 2000× with 400 DPI and was enlarged four times on screen. Even from this low magnification image, it is possible to estimate apparent pearlite spacings of 70–80 nm with a ±20 nm resolution. A higher magnification and better digital recordings are possible today, leading to the above cited 5–10 nm target value. Comparable measurements using TEM or atom probe are practically impossible considering time and cost.

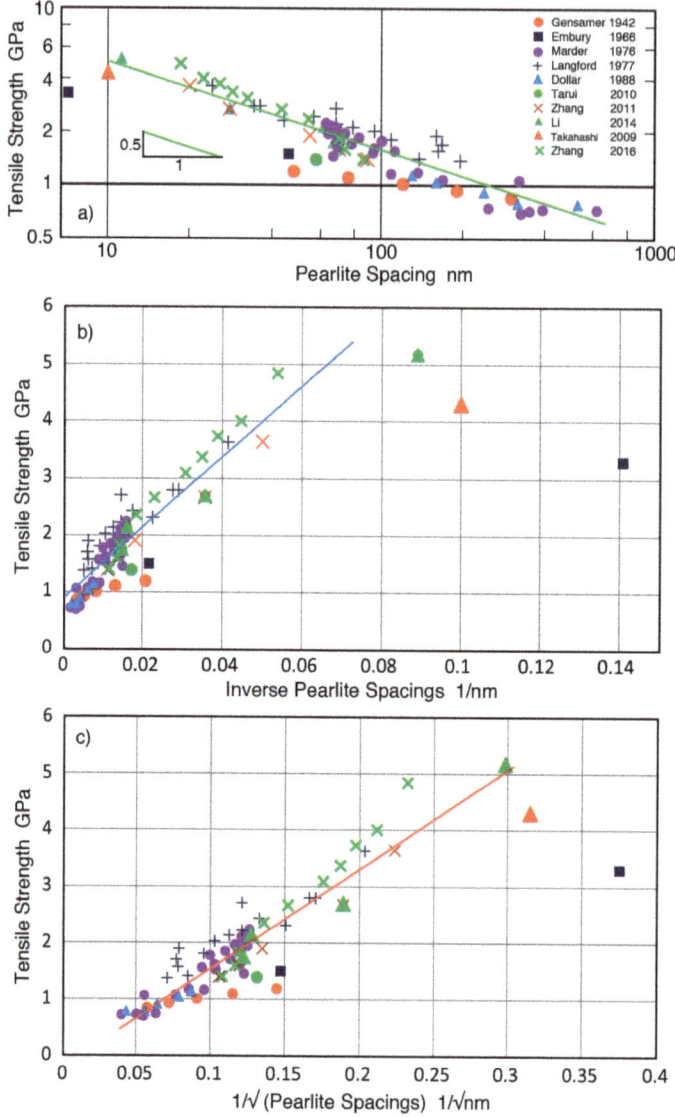

Figure 7. Tensile strength versus length parameters using data from [3,29,30,32,34,40,49,50,73,75]. (**a**) Pearlite spacings, in log-log scales. Data symbols are shown as insert. Green line is for a power-law fit with the slope of −0.5 giving $R^2 = 0.833$. (**b**) Inverse pearlite spacings. Same data symbols as in (**a**). Blue line is the regression line with $R^2 = 0.853$. (**c**) Inverse square-root of pearlite spacings. Same data symbols as in (**a**). Green line is the regression line with $R^2 = 0.868$.

5. Conclusions

This study was initiated to identify the suitability of using either the Hall-Petch or Griffith equation for describing the size effects of high strength pearlitic steels, which have been examined by many researchers. It is found that published size dependence data can be represented by both equations equally well. The strength versus pearlite spacing correlation also indicated statistical equivalence of the two relations, although improvements in data consistency are needed. A power-law equation was shown to be a form of the Hall-Petch equation, but even this can be approximated well using the Griffith equation with properly chosen constants. Consequently, the choice between two groups of theories that predict respective relationships must rely, for now, on the merit of theoretical developments and assumptions made [12,33,35,49,50].

Funding: This research received no external funding.

Acknowledgments: The author is grateful to Toshimi Tarui for extensive discussion and for finding valuable steel test data from sources commonly inaccessible and to Stephen Walley and Ron Armstrong for providing a copy of Percy article containing rare technical data from the late 19th century and for valuable comments.

Conflicts of Interest: The author declares no conflict of interest.

Appendix A. An Approximate Method of Estimating Weibull Shape Parameter, m

It has been shown that the degradation of high strength cable wires of high C pearlitic steels can be characterized by Weibull shape parameter or Weibull modulus [26]. New wires are expected to have m values of more than 70 since $m = 70.4$ was found for wires of stage 2 corrosion damage. Actual Weibull analysis of high strength cable wires is apparently unavailable in the literature. Perry [78] reported old cable wires from the Williamsburg Bridge in New York (built in 1903) using wires removed during its rehabilitation project. A set of 160 tensile tests was analyzed and m was found to be 16.0 [4] (in the original report, a computational error was made and m was reported as 2.303 times higher than the correct value.) Another data set was given by Percy [11] and m was found to be 13.7, using the same method found in [4]. Weibull plots of these two cases are given in Figure A1 with the slope of m, which are the only two available in the literature. Usually, values of the tensile strength of high strength cable wires are reported collectively with an average value plus standard deviation. Assuming the fracture behavior follows the Weibull statistics, it is possible to estimate an approximate value of m. This is done by getting stress values in Equation (4) by supplying a set of P and m values. By comparing the average and standard deviation of the stress values with the corresponding observed values, m value can be estimated by iteration.

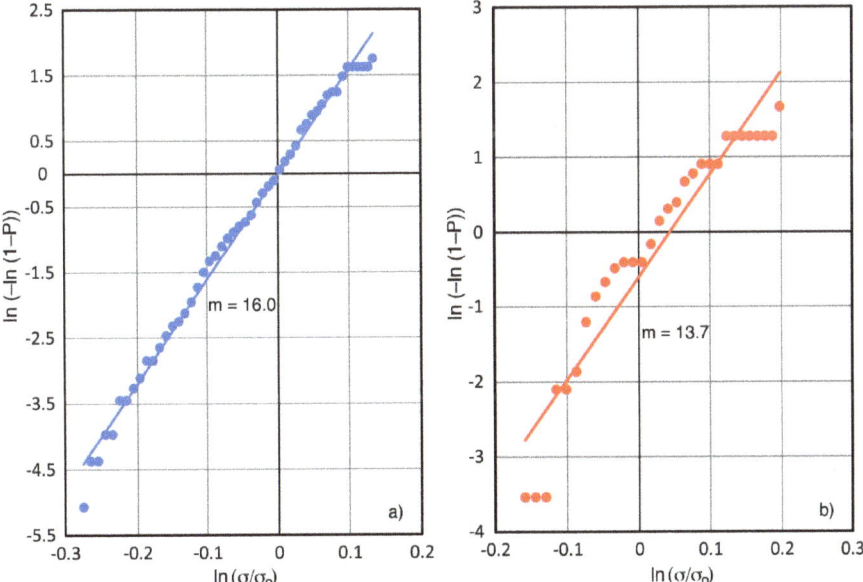

Figure A1. Weibull plots for steel wires. (**a**) Williamsburg Bridge (1903), data from [78], (**b**) Percy (1886), data from [11].

Suppose the average tensile strength (TS) and its standard deviation (SD) are given. When SD is a few % of TS, m is expected to be higher than 50, while 5–10% level indicates m to be between 10 and 30. Pick a trial value of m and use TS as the scale parameter, σ_o. In addition, pick a large number, N (e.g., 10,000). Using Excel, Col A should have 0.0001 (= 1/N) to 1. Col B is set to = LN(−LN(1 − Ann)), nn being 1 to 10,000. This is to calculate values of $m \cdot \ln(\sigma/\sigma_o)$. Col C is set to = $\sigma_o \cdot$EXP(Bnn/m). This gives values of σ that corresponds to P values in Col A. Compute <σ> = AVERAGE(C1:C10,000) and <sd> = STDEV(C1:C10,000). Compare <σ> and <sd> to TS and SD. After several trials, these two calculated values should approach TS and SD. At this stage, also change the input σ_o value, by increasing TS with a factor F given by

$$F = 1 + 0.276\, m^{-0.776}, \quad (9)$$

This empirical factor that depends on m is needed because σ_o and the average value or <σ>, given by Equation (5), differ. The calculated average of the σ value, <σ>, quickly converges to TS, and when the STDEV value, or <sd>, becomes close to SD, this m value is taken as an estimate. This procedure was confirmed to provide valid m values for the data with known m, TS, and SD values. Five known cases from [4,26] matched. The Percy's 1886 data mentioned above showed a slightly higher m value of 15.3, which is likely due to large scatter in the original data. Results are given in Table A1. Three additional cases without m data are also examined. One data set was from the Mid-Hudson Bridge main cable [79]. The wires had SD of 4.7% of TS and showed stage 2 and 3 level corrosion and gave $m = 30$, comparable to the data in [26]. Two other cable wire data for TS and SD came from Japan, through the search conducted by Toshimi Tarui of NSSM. One set was the actual test data for the main cables of the Bisan Seto Bridge (1100 m main span, completed 1988) [80]. The sample count was 38,470 and the present method yielded an estimated $m = 110$. The second set was a supplemental data of a Japanese Standard for bridge cable wires (5-mm galvanized steel wires) [81]. This wire test data with sample count of 45 resulted in $m = 124$. These two estimated m values confirm the high quality being achieved for the fabrication of suspension bridge cable wires.

Table A1. Input data and results of Weibull analysis.

m	m est	TS GPa	SD GPa	TS est GPa	SD est GPa	σ_o GPa	σ_o/TS	Sample Counts	Ref.
9.1	9.1	1.3830	0.1815	1.3830	0.1820	1.4600	1.0557	15	[26]
13.7	15.3	1.0920	0.0878	1.0920	0.0876	1.1300	1.0348	35	[11]
16.0	16.0	1.4990	0.1130	1.5000	0.1150	1.5500	1.0340	160	[4]
N/A	30.0	1.6390	0.0773	1.6400	0.0680	1.6700	1.0189	N/A	[79]
33.4	33.4	1.5950	0.0600	1.5950	0.0600	1.6215	1.0166	15	[26]
52.4	52.4	1.6280	0.0393	1.6270	0.0393	1.6450	1.0104	15	[26]
70.6	70.6	1.6490	0.0297	1.6490	0.0296	1.6620	1.0079	20	[26]
N/A	110	1.6530	0.0192	1.6540	0.0192	1.6630	1.0060	38470	[80]
N/A	124	1.6600	0.0171	1.6600	0.0170	1.6680	1.0048	45	[81]

Known values are in green, while those in red are estimated by the present method. N/A: Not available, est: estimated; Ref.: reference number.

Omitted above in the steps for Weibull parameter calculation is another variable or $\ln(\sigma/\sigma_o)$. For this, add Col D and set = Bnn/m. The Weibull plot can then be obtained by plotting Col D and Col B. In the present case, it produces a straight line with the slope of m. Another useful plot for examining the data characteristics is to plot stress (Col C) versus P(Col A), yielding the cumulative probability distribution curve. This shows a skewed S-shaped curve. The present method can estimate the Weibull modulus for common mechanical property data when it was obtained with statistically valid sample counts. Usually, this may suggest counts of more than 30–50. However, only 15 to 20 samples were used in [26] and still worthy m values were determined. Thus, the sample counts needed are within normal engineering practice.

References

1. Sluszka, P. *Studies on the Longevity of Suspension Bridge Cables*; Transportation Research Record 1290; National Academy of Sciences: Washington, DC, USA, 1990; pp. 272–278.
2. Nippon Steel Sumitomo Metal Corp. Current Conditions of the Development and Practical Application of Wire Rods for Bridge Cable Use That Have the Highest Level of Strength Worldwide. Available online: http://www.nssmc.com/en/steelinc/information/20150423_100.html (accessed on 18 April 2018).
3. Li, Y.J.; Raabe, D.; Herbig, M.; Choi, P.; Goto, S.; Kostka, A. Segregation stabilizes nanocrystalline bulk steel with near theoretical strength. *Phys. Rev. Lett.* **2014**, *113*, 106104. [CrossRef] [PubMed]
4. Ono, K. Structural materials: Metallurgy of bridges. In *Metallurgical Design and Industry, Prehistory to the Space Age*; Kaufman, B., Briant, C.L., Eds.; Springer: Cham, Switzerland, 2018; pp. 193–269.
5. Plowden, D. *Bridges, the Spans of North America*; Viking Press: New York, NY, USA, 1974; p. 328.
6. Pan, H.X. *Famous Ancient Bridges of China*; Shanghai Cultural Publishing House: Shanghai, China, 1985. (In Chinese)
7. Pan, H.X. Study of Yunnan Lanjin and Ji Hong bridges. *J. Tongji Univ.* **1981**, *1*, 108–116. (In Chinese)
8. Pope, T. *A Treatise on Bridge Architecture, in Which the Superior Advantages of the Flying Pendent Lever Bridge Are Fully Proved*; Self-published: New York, NY, USA, 1811; p. 279.
9. Peters, T.F. *Transitions in Engineering*; Birkhauser Verlag: Basel, Switzerland, 1987; p. 244.
10. Dodge, A. *Pianos and Their Makers*; Covina Publ. Co.: Covina, CA, USA, 1911; pp. 123–126.
11. Percy, J. On steel wire of high strength. *J. Iron Steel Inst.* **1886**, *29*, 62–80.
12. Armstrong, R.W. Plasticity: Grain Size Effects III. In *Reference Module in Materials Science and Engineering*; Hashmi, S., Ed.; Elsevier: New York, NY, USA, 2016; pp. 1–23.
13. Wright, R.N. *Wire Technology: Process Engineering and Metallurgy*; Butterworth-Heinemann: New York, NY, USA, 2014; p. 340.
14. Griffith, A.A. The phenomenon of rupture and flow in solids. *Philos. Trans. R. Soc. Lond.* **1921**, *A221*, 163–198. [CrossRef]
15. Ochiai, I.; Nishida, S.; Ohba, H.; Kawana, A. Application of Hypereutectoid Steel for Development of High Strength Steel Wire. *Tetsu-to-Hagane* **1993**, *79*, 1101–1107. [CrossRef]
16. Klingsporn, P.E. *Characterization of Optical Fiber Strength under Applied Tensile Stress and Bending Stress*; KCP-613-6655; Honeywell: Kansas City, MO, USA, 2011; 44p.
17. Karmarsch, I. *Mittheilungen des gew Ver. für Hannover*; Hahnschehof Publ: Hannover, Germany, 1858; pp. 138–155.
18. Rubenstein, L.S. *Effects of Size on Tensile Strength of Fine Polycrystalline Nickel Wires*; NASA-TN-4884; National Aeronautics and Space Administration: Washington, DC, USA, 1968; 25p.
19. Riesch, J.; Feichtmayer, A.; Fuhr, M.; Almanstötter, J.; Coenen, J.W.; Gietl, H.; Höschen, T.; Linsmeier, C.; Neu, R. Tensile behaviour of drawn tungsten wire used in tungsten fibre-reinforced tungsten composites. *Phys. Scr.* **2017**, *T170*, 14032. [CrossRef]
20. *Report of the Tests of Metals and Other Materials for Industrial Purposes*; US Testing Machine: Watertown Arsenal, Boston, MA, USA, 1895.
21. Zhu, Y.T.; Butt, D.P.; Taylor, S.T.; Lowe, T.C. Evaluation of modified Weibull distribution for describing the strength of ceramic fibers and whiskers with varying diameters. *J. Test. Eval.* **1998**, *26*, 144–150. [CrossRef]
22. Weibull, W. A statistical distribution function of wide applicability. *J. Appl. Mech.* **1951**, *73*, 293–297.
23. Otto, W.H. Relationship of tensile strength of glass fibers to diameter. *J. Am. Chem. Soc.* **1955**, *39*, 122–124. [CrossRef]
24. Flores, O.; Bordia, R.K.; Bernard, S.; Uhlemann, T.; Krenkel, W.; Motz, G. Processing and characterization of large diameter ceramic SiCN monofilaments from commercial oligosilazanes. *RSC Adv.* **2015**, *5*, 107001. [CrossRef]
25. Catangiu, A.; Ungureanu, D.N.; Despa, V. Data scattering in strength measurement of steels and glass/epoxy composite. *Mat. Mech.* **2017**, *15*, 11–16. [CrossRef]
26. Mayrbaurl, R.M.; Camo, S. *Guidelines for Inspection and Strength Evaluation of Suspension Bridge Parallel Wire Cables*; Report 534; National Cooperative Highway Research Program: Washington, DC, USA, 2004; p. 274.

27. Waugh, A.R.; Paetke, S.; Edmonds, D.V. A study of segregation to the dislocation substructure in patented steel wire using atom-probe techniques. *Metallography* **1981**, *14*, 237–251. [CrossRef]
28. Hong, M.H.; Reynolds, W.T., Jr.; Tarui, T.; Hono, K. Atom probe and transmission electron microscopy investigations of heavily drawn pearlitic steel wire. *Metall. Mater. Trans. A* **1999**, *30A*, 717–727. [CrossRef]
29. Takahashi, J.; Tarui, T.; Kawakami, K. Three-dimensional atom probe analysis of heavily drawn steel wires by probing perpendicular to the pearlitic lamellae. *Ultramicroscopy* **2009**, *109*, 193–199. [CrossRef] [PubMed]
30. Embury, J.D.; Fisher, R.M. The structure and properties of drawn pearlite. *Acta Metall.* **1966**, *14*, 147–152. [CrossRef]
31. Langford, G. A study of the deformation of patented steel wire. *Metall. Trans.* **1970**, *1*, 465–477. [CrossRef]
32. Langford, G. Deformation of pearlite. *Metall. Trans. A* **1977**, *8*, 861–875. [CrossRef]
33. Langford, G.; Cohen, M. Calculation of cell-size strengthening of wire drawn iron. *Metall. Trans.* **1970**, *1*, 1478–1480. [CrossRef]
34. Marder, A.R.; Bramfitt, B.L. The effects of morphology on the strength of pearlite. *Metall. Trans. A* **1976**, *7A*, 365–372. [CrossRef]
35. Borchers, C.; Kirchheim, R. Cold-drawn pearlitic steel wires. *Prog. Mater. Sci.* **2016**, *82*, 405–444. [CrossRef]
36. Tashiro, H. The challenge for maximum tensile strength steel cord. *Nippon Steel Tech. Rep.* **1999**, *80*, 6–8.
37. Kanetsuki, Y.; Ibaraki, N.; Ashida, S. Effect of cobalt addition on transformation behavior and drawability of hypereutectoid steel wire. *ISIJ Int.* **1991**, *31*, 304–311. [CrossRef]
38. Choi, H.C.; Park, K.T. The effect of C content on the Hall-Petch parameters in the cold-drawn hypereutectoid steels. *Scr. Mater.* **1996**, *34*, 857–862. [CrossRef]
39. Maruyama, N.; Tarui, T.; Tashiro, H. Atom probe study on the ductility of drawn pearlitic steels. *Scr. Mater.* **2002**, *46*, 599–603. [CrossRef]
40. Zhang, X.D.; Godfrey, A.; Huang, X.X.; Hansen, N.; Liu, Q. Microstructure and strengthening mechanisms in cold-drawn pearlitic steel wire. *Acta Mater.* **2011**, *59*, 3422–3430. [CrossRef]
41. Nam, W.J.; Bae, C.M. Void initiation and microstructural changes during wire drawing of pearlitic steels. *Mater. Sci. Eng.* **1995**, *203*, 278–285. [CrossRef]
42. Zelin, M. Microstructure evolution in pearlitic steels during wire drawing. *Acta Mater.* **2002**, *50*, 4431–4447. [CrossRef]
43. Kim, D.K.; Shemenski, R.M. Alloy Steel Tire Cord and Its Heat Treatment Process. US Patent 5,167,727, 1 December 1992.
44. Goto, S.; Kirchheim, R.; Al-Kassab, T.; Borchers, C. Application of cold drawn lamellar microstructure for developing ultra-high strength wires. *Trans. Nonferr. Met. Soc. China* **2007**, *17*, 1129–1138. [CrossRef]
45. Pepe, J.J. Deformation structure and tensile fracture characteristics of a cold worked 1080 pearlitic steel. *Metall. Trans.* **1973**, *4*, 2455–2460. [CrossRef]
46. Buono, V.T.L.; Gonzalez, B.L.; Lima, T.M.; Andrade, M.S. Measurement of fine pearlite interlamellar spacing by atomic force microscopy. *J. Mater. Sci.* **1997**, *32*, 1005–1008. [CrossRef]
47. Yamakoshi, N.; Nakamura, Y.; Kaneda, T. Development of high strength steel wires. *R D Kobe Steel Tech. Rep.* **1973**, *23*, 20–27.
48. Makii, K.; Tarui, T.; Tsuzaki, K. *Improvements in the Strength and Reliability of Steels*; The Iron and Steel Institute of Japan: Tokyo, Japan, 1997; p. 3.
49. Tarui, T. On Enhancement of Strength and Ductility of High Carbon Steel Wires. Ph.D. Thesis, Tokyo Institute of Technology, Tokyo, Japan, 2010.
50. Zhang, X.; Hansen, N.; Godfrey, A.; Huang, X. Dislocation-based plasticity and strengthening mechanisms in sub-20 nm lamellar structures in pearlitic steel wire. *Acta Mater.* **2016**, *114*, 176–183. [CrossRef]
51. Glaesemann, G.S. Optical fiber failure probability predictions from long-length strength distributions. In Proceedings of the 40th International Wire and Cable Symposium, St. Louis, MO, USA, 18–21 November 1991; pp. 819–825.
52. Tagawa, T.; Miyata, T. Size effect on tensile strength of carbon fibers. *Mater. Sci. Eng. A* **1997**, *A238*, 336–342. [CrossRef]
53. Nakagawa, S.; Morimoto, T.; Ogihara, S. The size effect of SiC fiber strength on the gauge length. In Proceedings of the 24th international congress of aeronautical sciences, Yokohama, Japan, 29 August–3 September 2004.

54. Wagner, H.D. Stochastic concepts in the study of size effects in the mechanical strength of highly oriented polymeric materials. *J. Polym. Sci. Polym. Phys.* **1989**, *27*, 115–149. [CrossRef]
55. Schwartz, P.; Netravali, A.; Sembach, S. Effects of strain rate and gauge length on the failure of ultra-high strength polyethylene fibers. *Text. Res. J.* **1986**, *56*, 502–508. [CrossRef]
56. Smook, J.; Hamersma, W.; Pennings, A.J. The fracture process of ultra-high strength polyethylene fibres. *J. Mater. Sci.* **1984**, *19*, 1359–1373. [CrossRef]
57. Wagner, H.D.; Steenbakkers, L.W. Stochastic strength and size effect in ultra-high strength polyethylene fibres. *Philos. Mag. Lett.* **1989**, *59*, 77–85. [CrossRef]
58. Roffey, P. The fracture mechanisms of main cable wires from the forth road suspension. *Eng. Fail. Anal.* **2013**, *31*, 430–441. [CrossRef]
59. Morgado, T.L.M.; Sousa e Brito, A. A failure analysis study of a prestressed steel cable of a suspension bridge. *Case Studies Constr. Mater.* **2015**, *3*, 40–47. [CrossRef]
60. Mahmoud, K.M. Fracture strength for a high strength steel bridge cable wire with a surface crack. *Theor. Appl. Fract. Mech.* **2015**, *48*, 152–160. [CrossRef]
61. Chida, T.; Hagihara, Y.; Akiyama, E.; Iwanaga, K.; Takagi, S.; Hayakawa, M.; Ohishi, H.; Hirakami, D.; Tarui, T. Comparison of constant load, SSRT and CSRT methods for hydrogen embrittlement evaluation using round bar specimens of high strength steels. *ISIJ Int.* **2016**, *56*, 1268–1275. [CrossRef]
62. Mahmoud, K.; Hindshaw, W.; McCulloch, R. *Management Strategies for Suspension Bridge Main Cables*; Asset Management of Bridges; Mahmoud, K., Ed.; CRC Press: Leiden, The Netherland, 2017; pp. 3–12.
63. Bennett, J.A.; Mindlin, H. Metallurgical aspects of the failure of the Point Pleasant Bridge. *J. Test. Eval.* **1973**, *1*, 152–161.
64. Tarui, T.; Maruyama, N.; Eguchi, T.; Konno, S. High strength galvanized steel wire for bridge cables. *Struct. Eng. Int.* **2002**, *12*, 209–213. [CrossRef]
65. Nakamura, S.; Suzumura, K.; Tarui, T. Mechanical properties and remaining strength of corroded bridge wires. *Struct. Eng. Int.* **2004**, *14*, 51–54. [CrossRef]
66. Nakamura, S.; Suzumura, K. Hydrogen embrittlement and corrosion fatigue of corroded bridge wires. *J. Constr. Steel Res.* **2009**, *65*, 270–275. [CrossRef]
67. Fang, F.; Zhao, Y.; Liu, P.; Zhou, L.; Hu, X.J.; Zhou, X.; Xie, Z.H. Deformation of cementite in cold drawn pearlitic steel wire. *Mater. Sci. Eng. A* **2014**, *608*, 11–15. [CrossRef]
68. Takahashi, J.; Kosaka, M.; Kawakami, K.; Tarui, T. Change in carbon state by low-temperature aging in heavily drawn pearlitic steel wires. *Acta Mater.* **2012**, *60*, 387–395. [CrossRef]
69. Ono, K.; Sommer, A.W. Peierls-Nabarro hardening in the presence of point obstacles. *Metall. Trans.* **1970**, *1*, 877–884. [CrossRef]
70. Das, A. Calculation of ductility from pearlite microstructure. *Mater. Sci. Technol.* **2018**, *34*, 1046–1063. [CrossRef]
71. Ohtsu, M.; Ono, K. Pattern recognition analysis of magnetomechanical acoustic emission signals. *J. Acoust. Emiss.* **1984**, *3*, 69–80.
72. Gensamer, M.; Pearsall, E.B.; Smith, G.V. The mechanical properties of the isothermal decomposition products of austenite. *Trans. ASM* **1940**, *28*, 380–398.
73. Gensamer, M.; Pearsall, E.B.; Pellini, W.S.; Low, J.R., Jr. The tensile properties of pearlite, bainite, and spheroidite. *Metall. Microstruct. Anal.* **2012**, *1*, 171–189, Reprinted from *Trans. ASM* **1942**, *30*, 983–1019. [CrossRef]
74. Ridley, N. A review of the data on the interlaminar spacings of pearlite. *Metall. Trans. A* **1984**, *15A*, 1019–1036. [CrossRef]
75. Dollar, M.; Bernstein, I.M.; Thompson, A.W. Influence of deformation substructure on flow and fracture of fully pearlitic steel. *Acta Metall.* **1988**, *36*, 311–320. [CrossRef]
76. Vander Voort, G.F.; Roosz, A. Measurement of the interlamellar spacing of pearlite. *Metallography* **1984**, *17*, 1–17. [CrossRef]
77. Tashiro, H.; Sato, H. Effect of alloying elements on the lamellar spacing and the degree of regularity of pearlite in eutectoid steel. *J. Jpn. Inst. Met.* **1991**, *55*, 1078–1085. [CrossRef]
78. Perry, R.J. Estimating strength of the Williamsburg Bridge suspension cables. *Am. Statistician* **1998**, *52*, 211–217.
79. Mahmoud, K.M. *BTC Method for Evaluation of Remaining Strength and Service Life of Bridges*; NYSDOT Report C-07-11; Bridge Technology Consulting: New York, NY, USA, 2011; pp. 21–24.

80. Japan Bridge Engineering Center. *High Strength (1.8 GPa Class) Galvanized Steel Wires for Bridge Cables*; Report on Bridge Cable Design Methods; JBEC: Tokyo, Japan, 1988.
81. Japan Society. *Steel Construction and Nippon Steel Sumitomo Metals, Standard for structural cable materials*; Japanese Industrial Standards II, 03,04,05,06,11-1994; Japanese Industrial Standards Committee: Tokyo, Japan, 1994.

© 2019 by the author. Licensee MDPI, Basel, Switzerland. This article is an open access article distributed under the terms and conditions of the Creative Commons Attribution (CC BY) license (http://creativecommons.org/licenses/by/4.0/).

Review

Dislocation Emission and Crack Dislocation Interactions

Chandra S. Pande [1] and Ramasis Goswami [2,*]

1. Volunteer Emeritus, Naval Research Laboratory, Washington, DC 20375, USA; chandrasp22@gmail.com
2. Materials Science and Technology Division, Naval Research Laboratory, Washington, DC 20375, USA
* Correspondence: Ramasis.Goswami@nrl.navy.mil

Received: 3 January 2020; Accepted: 19 March 2020; Published: 3 April 2020

Abstract: An understanding of the crack initiation and crack growth in metals spanning the entire spectrum of conventional and advanced has long been a major scientific challenge. It is known that dislocations are involved both in the initiation and propagation of cracks in metals and alloys. In this review, we first describe the experimental observations of dislocation emission from cracks under stress. Then the role played by these dislocations in fatigue and fracture is considered at a fundamental level by considering the interactions of crack and dislocations emitted from the crack. We obtain precise expression for the equilibrium positions of dislocations in an array ahead of crack tip. We estimate important parameters, such as plastic zone size, dislocation free zone and dislocation stress intensity factor for the analysis of crack propagation. Finally, we describe very recent novel and significant results, such as residual stresses and relatively large lattice rotations across a number of grains in front of the crack that accompanies fatigue process.

Keywords: crack tip dislocations; TEM; grain rotation; fatigue; dislocation configurations; residual stress

1. Introduction

The prediction of the fatigue properties of structural materials is rightly recognized as one of the most important problems in engineering. Previous works suggest that fatigue is an very complex phenomenon primarily because of the large number of variables spanning the aspects of microstructure, alloy chemistry, processing treatment, intrinsic microstructural effects, and test variables. The frequency of failure mechanisms of aircraft components as a result of corrosion, fatigue, overload, high temperature corrosion, corrosion fatigue and wear/abrasion/erosion is 16%, 55%, 14%, 2%, 7% and 6%, respectively [1]. In fatigue phenomenon, the unsolved fundamental questions are: (i) how cracks are created? (ii) Where do the dislocations originate? (iii) How do the cracks interact with mobile and immobile dislocations? (iv) How cracks are able to grow at loads far less than that needed for fracture? and (v) What are the effects microstructural features, such as: voids, interfaces, grain boundaries, and second-phase particles? We will be able to predict the fatigue behavior in a variety of materials and structures, once these questions have been addressed.

Basinski and Basinski [2] reported that cracks nucleate at persistent slip bands (PSBs) that are generated immediately prior to fracture initiation. According to Mott [3], vacancies are generated immediately below the surface and eventually coalesce to form fine microscopic cracks. This idea has been extended by Antonopoulos et al. [4]. They proposed a model based on vacancy dipoles, which develop in the PSBs. Essmann et al. [5] employed similar ideas for the nucleation of fine microscopic cracks. However, Neumann [6] developed and put forward a model based on an activation of two operating slip systems. It, therefore, appears that the crack nucleation process has been fairly well understood for a wide spectrum of materials.

In the next section, we describe and discuss the emission of dislocations from the cracks and provide some direct evidence for such emission. In Section 3, we discuss in detail used to predict various features of crack process. This forms the main part of this paper. In Section 4, we some other features of the fatigue usually not considered that is, to determine characteristics of lattice rotation and residual stresses around and in front of the crack , and relate it to fatigue plastic zone and dislocation configurations, in high stacking fault energy materials, particularly in Al alloys. Finally, in Section 5, we provide some discussion and concluding remarks. The objective of this brief review is:

(i) To establish that dislocation emission from cracks plays a major role in the fatigue process.
(ii) To point out the methodologies and recent progresses in analytical formulation of the dislocation interactions, and the role they play in fatigue and fatigue processes.
(iii) To present some very recent results pertaining to the changes in grain morphology, orientations and residual stresses as the materials undergo fatigue.

2. Dislocations Emission From Cracks

Significant efforts have been carried out to examine the crack-tip deformation behavior in metals and alloys using in-situ tensile deformation in an electron microscope. It has been reported that deformation occurs mostly by the emission of dislocations from the crack tip. The propagation of cracks has been correlated with the behavior of dislocations ahead of the crack tip. The process of dislocation emission from the crack tip have been examined theoretically by several authors, considering the elastic interaction between a crack and a dislocation. Ohr et al. [7–9] have observed dislocation emissions and the distribution of dislocations in the plastic zone during in-situ tensile deformation in an electron microscope. They reported a region immediately ahead of the crack tip to be free of dislocations and called this region as dislocation free zone (DFZ). Similar observations have been made by Park and coworkers [10].

Transmission electron microscopy (TEM) was employed to investigate the spatial configuration of dislocations emitted from crack tip in Cu and Al alloys. Figure 1a shows the emission of dislocation from a crack in (111) oriented single crystalline Cu. In this case, the crack lies on one of the (111) planes, and the dislocations are long and curved (see Figure 1b). Figure 2a is a bright-field TEM image showing the dislocations emitted from a crack tip for Al and a dislocation free zone (DFZ) of ≈ 1.35 μm in length ahead of the crack tip. A total of ≈ 90 dislocations was observed to emitted from the crack tip [11]. These dislocations were then migrated up to 3.85 μm from the crack tip. The plastic zone was found to be elliptical, which is in agreement with the discrete dislocation model [12], and the distribution of dislocations is approximately in the form of inverse pile-up configurations, which is consistent with the previous observations for stainless steel, Cu, Ni, Al and Mo [13]. We have examined the dislocations under different tilt conditions from 10 to 45° to investigate the dislocation configurations (see Figure 2b–e) in the plastic zone, and determine the Burgers vector of the dislocations using the $g.b = 0$ criterion. Figure 3a shows most of the dislocations with g = $\bar{1}1\bar{1}$ are invisible. Most dislocations are, however, visible with g = $1\bar{1}\bar{1}$ and g = 002 as shown in Figure 3b. It suggests that the most dominant dislocations have Burgers vector, $b = \frac{a}{2}[011]$. For the case of thin foil, these dislocations around the crack tip were emitted to accommodate mode-III stress intensity. Note that usually in mode-III loading, most dislocations are of screw type, and in this case these dislocations lie on one parallel set of 111 planes. The estimated mean position of dislocations is in reasonably good agreement with models of crack-dislocation configuration based on a continuum distribution dislocations. These models, however, do not accurately predict the number of dislocations emitted by the crack tip.

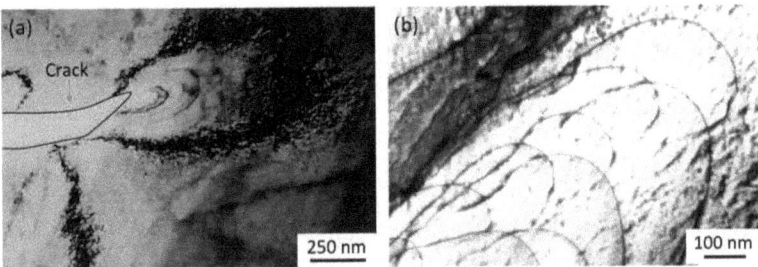

Figure 1. (a,b) A bright-field transmission electron microscopy (TEM) image close to the [111] zone in Cu showing curved dislocations ahead of a sharp crack

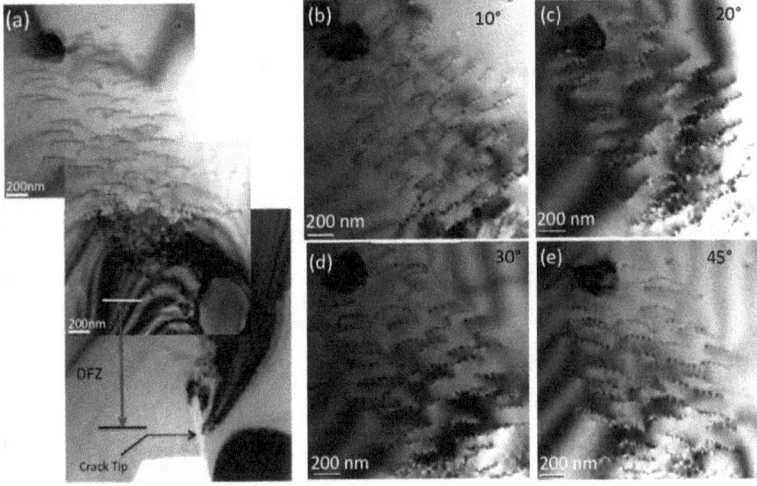

Figure 2. (a) A TEM image showing dislocations ahead of a sharp crack and a dislocation free zone (DFZ). (b–e) Dislocations configurations at different tilt angles of 10, 20, 30 and 45°, respectively.

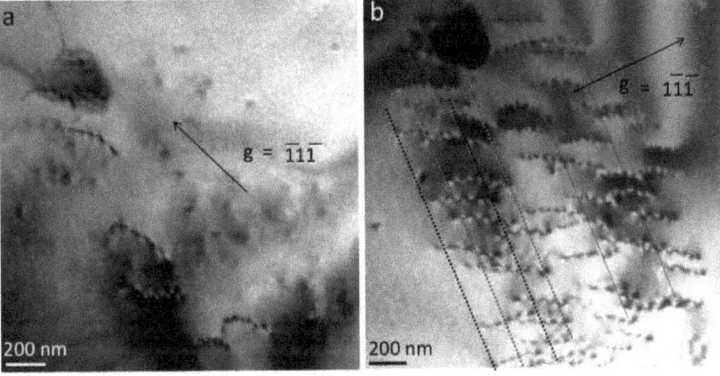

Figure 3. (a,b) Two beam bright field images showing dislocations with g = $\bar{1}1\bar{1}$ and g = $1\bar{1}\bar{1}$, respectively.

3. Crack-Dislocation Interaction(s)

To understand of the fatigue phenomenon at the microscopic level, an investigation of crack-dislocation interactions [14] is needed. In this regard, the most interesting work has been the development of discrete modeling approach by Pippan et al. [15–18], Deshpande et al. [19], and Mastorakos and Zbib [20]. They established that it is essential to concentrate on the initial stages of fatigue process. In addition, to better understand the existence of a threshold in the fatigue, an understanding of the stresses both at and near the crack tip, is required. The discrete dislocation studies may deliver the changes in both the stresses and resultant displacements during cyclic loading, and this model can convincingly show the crack propagation mechanism, which is appropriate depending on the nature of loading. Note that most models related to the fatigue phenomenon are two-dimensional in nature. Mastorakas and Zbib [20] employed a three-dimensional (3D) model, which is more realistic to obtain a better understanding of the fatigue phenomenon. However, it does not use of discrete dislocations concept. Several simulations using the discrete nature of dislocations have been done before [18,21–34], which suggest that the fatigue threshold behavior can be related to the discrete nature of plastic deformation.

As discussed before, considerable efforts have been made to experimentally investigate the crack-tip deformation behavior using TEM [35–38]. All of these experimental observations show that the crack tip acts as a possible source of dislocations. As a result the dislocation emission starts from the crack tip, which is consistent with the fact that the crack tip is associated with the highest stress. Such crack-tip dislocation emission process, on the other hand, have been theoretically [13,39–43] investigated by several authors. Several dislocation models [13,39–45] have been proposed to understand the initiation and propagation of cracks.

The initial model for the monotonically loaded cracks was first given by Bilby, Cottrell and Swinden (BCS) [44], which considers a finite crack in an infinite isotropic elastic medium (Figure 4). However, the BCS model does not consider for the existence of the DFZ in front of the crack tip, which has been experimentally observed. The BCS has been applied by Weertman [46] to model fatigue crack growth. This model has been modified by Chang and Ohr [13], Majumdar and Burns [39,40] and Weertman et al. [46–49]. Majumdar and Burns [39,40] model considers a crack tip with a pile-up of screw dislocation, and a dislocation free zone immediately after the crack. This model assumes continuous distribution of dislocations in the plastic zone. Lin and Thomson (LT) [45] considered two symmetrically inclined slip planes stemming from a semi-infinite crack tip, and they replaced the dislocation arrays by a superdislocation and examined the emission of dislocation. In the next section, we discuss another crack-tip dislocation model by Pande, Masumura and Chou [50,51], which do not use the superdislocation approximation. In their model, Pande et al. provided an expression for the length of DFZ, which cannot be obtained by the Lin-Thomson model because of the unrealistic (not experimentally observed) approximation of superdislocation.

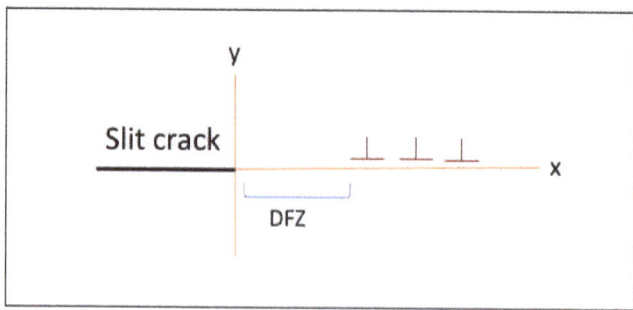

Figure 4. A schematic diagram showing the slit crack and the dislocation distribution.

Pande, Masamura and Chou Model

This model examined emission of dislocations on two symmetrically inclined slip planes from a semi-infinite crack tip in mode-I loading. The crack geometry is shown in Figure 5 for edge dislocations on two inclined slip planes. In this condition, the forces on the ith dislocation on the slip plane are due to (1) the crack tip stress field, (2) stress field due to other dislocations and the (3) image forces due to the crack surface. The sum of the forces, at equilibrium, is equal to the lattice friction force, F_{fric}. It can be written as [51];

$$F_{fric} = b\tau_{fric} = F_{image} + F_{cr} + F_d$$

$$F_{image} = -\frac{\mu b^2}{4\pi(1-\nu)r_i}$$

$$F_{cr} = \frac{bK_A}{\sqrt{2\pi r_i}} \sin\left(\frac{\theta}{2}\right) \cos^2\left(\frac{\theta}{2}\right) \tag{1}$$

$$F_d = \sum_{(k=1,k\neq i)}^{n} F_d^k,$$

where ν is Poissons ratio, K_A is the stress intensity factor, and F_{fric}, F_{image}, F_{cr} and F_d are the forces due to lattice friction, image, crack and dislocations, respectively.

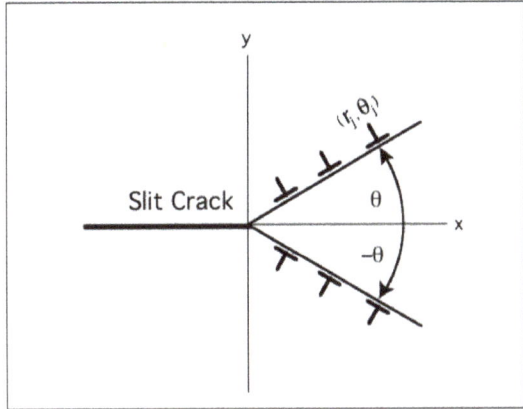

Figure 5. The diagram showing the dislocation distribution for $\theta = 70.54°$ in front of a slit crack.

The force due to other dislocations, F_d, can be estimated by the superposition of forces of the elastic field [52]. Zhang et al. [53] formulated the contribution of this force for an infinite slit crack and a single edge dislocation on an inclined slip plane using the complex potential [52]. The force acting on the ith dislocation due to the kth dislocation is given by the Peach-Koehler equation. It is written as;

$$F_d^k = b\sigma_{r\theta}^k \tag{2}$$

Thus, for the ith dislocation, the scaled force equilibrium condition is written as:

$$\tau_i^* = -\frac{1}{2\rho_i} + \frac{K^*}{\sqrt{\rho_i}} \sin\left(\frac{\theta}{2}\right) \cos^2\left(\frac{\theta}{2}\right) + \frac{1}{2} \sum_{(k=1,k\neq i)}^{n} f_d^k, \tag{3}$$

where $\tau_i^* = \frac{\tau_{fric}}{A}$, $f_d^k = \frac{\sigma_{r\theta}^k}{A}$, $\rho_i = \frac{r_i}{b}$, $A = \frac{\mu}{2\pi(1-\nu)}$ and $K^* = \frac{K_A}{A\sqrt{2\pi b}}$. For n dislocations, Equation (3) becomes a system of n equations, and the solution of these equations generates the equilibrium positions of dislocations, r_i, for a given K^* and τ_i^*. Utilizing the expressions given by Zhang et al. [53], the set of n × n non-linear simultaneous equations has been numerically solved.

One could use this model to estimate the size of the DFZ. The first and the last equilibrium positions are shown in Figure 6 as a function of the number of dislocations for $K^* = 10.0$ and $\tau_i^* = 0.1$. In this geometry, the slip plane inclination is $\theta = 70.54°$, and the plastic zone size is the distance from the crack tip to the last dislocation in the array. Similarly, the DFZ is the distance from the tip to the first dislocation. One could observe that the plastic zone size increases as the number of dislocations increases, while the DFZ decreases with the dislocation array. In this case, the greatest extent of the plastic zone size occurs when $\theta = 70.54°$, which corresponds to the maximum of the total applied stress field.

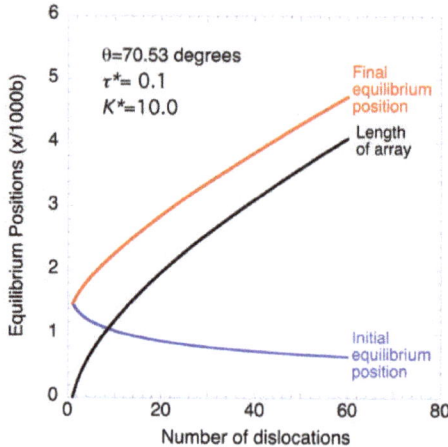

Figure 6. The diagram showing the equilibrium positions as a function of number of dislocations for $\theta = 70.54°$ and $\tau_i^* = 0.1$.

The centroid of the superdislocation in Figure 6 is calculated as a simple mean and is fairly linear over a large range of dislocations. Furthermore, it has been observed that the various zones decrease with decreasing K^*, for example, externally applied load. The elastic field at a crack tip is shielded ($b > 0$) or enhanced ($b < 0$), depending on the sign of the Burgers vector, which can be characterized by the stress intensity factor, K_D due to the dislocations, and have been given by Lin and Thomson [45] (see also Zhang et al. [53]). The stress intensity factor, K_D, is written as;

$$K_D = 6A\sqrt{\frac{\pi}{2}}\sin(\theta)\cos\left(\frac{\theta}{2}\right)\left[\frac{(n-1)}{\sqrt{\rho_{mean}}} + \frac{1}{\sqrt{\rho_l}}\right], \qquad (4)$$

where ρ_l is the position of superdislocation and n is number of dislocations. Using Equations (3) and (4), K_D has been plotted as a function of number of dislocation (see Figure 7). Here ρ_{mean} was considered as the arithmetic average of the dislocations positions. As K^* increases from 10 to 15, the externally applied tensile stress reduces the effect of shielding. Similarly, the decrease in friction stress, τ^* lowers the K_D. All these effects can influence the response to fatigue loads and the subsequent crack growth behavior. In some cases, the superdislocation computation can deviate significantly from the actual K_D, and the position of the superdislocation is somewhat arbitrary [50,51,54].

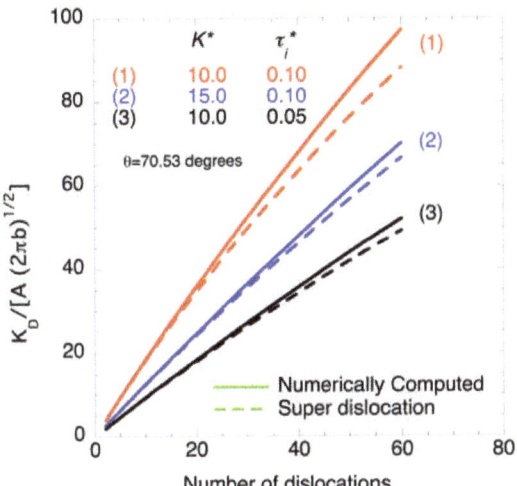

Figure 7. K_D as a function of number of dislocations. Note that the increase of K^* reduces the effect of shielding. The decrease of the friction stress lowers the K_D.

The distribution of dislocations is shown in Figure 8 for several values of n, $\theta = 70.54°$ and $\tau_i^* = 0.1$. As can be seen from the figure, the distribution is considerably different from a regular pile up of dislocation. The initial, final and equilibrium positions of dislocations [50,51,54] are shown in Figure 9, and it is clearly observed that the DFZ size is proportional to $\log n$, and the plastic zone is proportional to the number of dislocations in the array.

The lower limit for K^* and an equilibrium position, ρ_{min}, for a given τ^* and θ can be written as;

$$K^*_{min} = \frac{\sqrt{2\tau^*}}{\sin\left(\frac{\theta}{2}\right)\cos^2\left(\frac{\theta}{2}\right)} \tag{5}$$

$$\rho_{min} = \frac{1}{2\tau^*}.$$

For $\theta = 70.53°$ and $\tau^* = 0.1$, the above equation gives a minimum $K^* = 1.16$ and $\rho_{min} = 5$. This analysis shows a minimum value of K^* is required for the emission of dislocation from the crack. This minimum value of K^* could be associated with one of the two fatigue thresholds [55].

Here we briefly discuss the crack tip deformation behavior in mode-III loading. For $\theta = 0$, this reduces to analysis presented by Dai and Li [41]. The stress intensity factor increases with the increase in θ by a significant amount. For other parameters, such as length of pile up and DFZ, the estimates are similar to the values obtained by Dai and Li. Thus if the dislocation arrays are inclined, they might significantly affect the crack propagation, which is valid even if it is not exactly mode-III. These results are useful in brittle and ductile fracture [56].

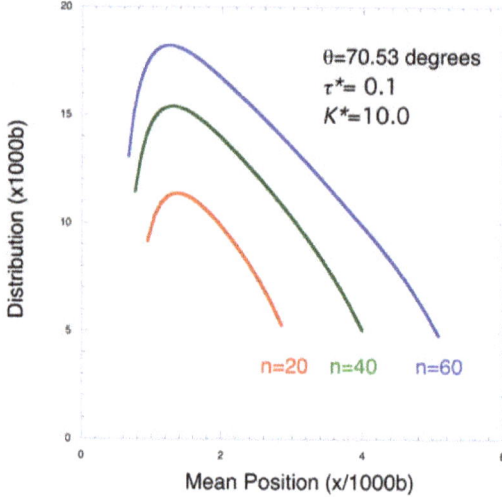

Figure 8. The diagram showing the first and last equilibrium positions as a function of the number of dislocations for $\theta = 70.54$ and $\tau_i^* = 0.1$.

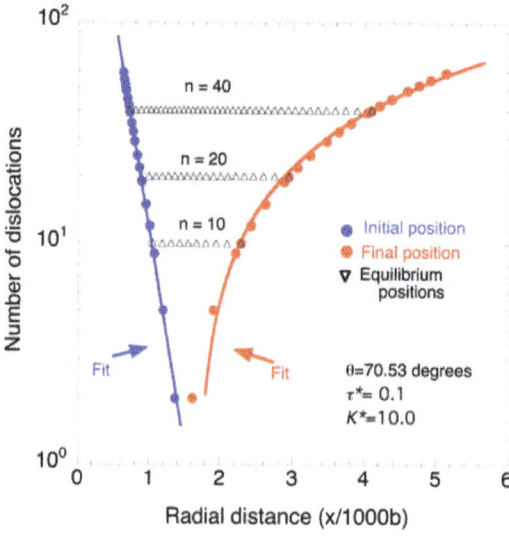

Figure 9. The diagram showing initial, final and equilibrium positions of dislocations.

4. Experimental Verification

The theoretical distribution, position and number of dislocation in the plastic zone have been compared here with the experimental observations of dislocations close to the crack. Here we use Majumdar and Burns' model to compare with the experimental observations as one can obtain the mean position and number of emitted screw dislocations for mode-III loadings. The experimental observations (see Figures 2 and 3) suggest they are mostly screw type. These experiments were carried out on thin foils using in-situ tensile stage in the transmission electron microscope, and in this case the loading in the tensile stage on the thin foil conforms to mode-III loading [38]. The distance of the mean position of dislocation, X_m, in front of the crack is given by Majumdar and Burns [39,40];

$$X_m = K\left(1 - \frac{c}{d}\right)^{0.5} E\left(\frac{c}{d}\right)^{0.5} c$$

$$K = \frac{\pi}{2} + \frac{m\pi}{8} + \frac{9\pi m^2}{128} + ...$$

$$E = \frac{\pi}{2} - \frac{m\pi}{8} - \frac{3\pi m^2}{128} - ...,$$

(6)

where K and E are the elliptic integrals, the c and d are the length of DFZ and the plastic zone size, respectively. The mean position, X_m, has been obtained from the experimentally measured parameters, c and d (see Figure 2). For Al, the experimentally observed values of c and d [11] are 1.35 µm and 3.85 µm, respectively, and the X_m turned out to be 2.15 µm. This suggests that the continuum approximation can adequately describe the plastic zone. One can now estimate the number of emitted dislocations and compare with the theory. The stress intensity factor (k_{III}) is given by;

$$k_{III} = 0.75\tau_{fric}\left(\frac{2c}{\pi}\right)^{0.5}\left[\ln\left(\frac{4d}{c}\right) + \frac{4}{3}\right].$$

(7)

The total stress intensity factor (K_{III}) is written as;

$$K_{III} = k_{III} + \sum_{j=1}^{N} k_{(j)}^{D},$$

(8)

where the second term, $\sum_{j=1}^{N} k_{(j)}^{D}$, is due to the contribution of N dislocations. Thus, the total stress intensity factor, K_{III}, can be written as [42];

$$K_{III} = 2\tau_{fric}\left(\frac{2d}{\pi}\right)^{0.5},$$

(9)

From Equation (9), the total number of crack tip dislocations, N, is given by;

$$N = \left(\frac{K_{III}}{\mu b}\right)\left(\frac{2d}{\pi}\right)^{0.5}$$

$$N = 4\frac{d}{\pi\mu b}\tau_{fric}.$$

(10)

Considering τ_{fric} = 20 MPa and μ = 27 GPa for Al, the number of emitted dislocations calculated using Equation (10) turned out to be 13. TEM observations, however, showed that the observed number of dislocations (see Figures 2 and 3) is approximately one order of magnitude greater than this estimation. Although the mean position of the dislocation distribution in the plastic zone can be predicted by the continuum dislocation models, they do not properly predict the total number of emitted dislocations from a crack tip. This discrepancy could result from the assumption that all dislocations lie on one plane is not a valid one. There are many other reasons also. The model itself is two dimensional and the dislocations are considered straight and infinite in length. One important fact observed in our experiments is that the crack has several sources of dislocations active at the same time. The predictions are thus merely quantitative. But such a large discrepancy need further experimental and theoretical investigations.

5. Effect of Dislocation Emissions

The emission of dislocations in fatigue results in significant lattice rotation and tensile residual stresses around the fatigue crack [57–60]. The lattice rotation may be related to the size of the plastic zone and the redistribution of dislocations and slip processes in front of crack. To study the plastic zone, the growth of the crack has been stopped at a different lengths. Fatigue tests were performed in vacuum (<6 × 10^{-6} Pa) background pressure at a cyclic load frequency of 10 Hz with a load ratio of 0.10 on compact tension (CT) specimens. Figure 10a shows the XRD patterns at different locations from the crack. One could observe the relative variations of 111 and 200 Al peaks (see Figure 10b) as a function of position from crack at either side of the crack for Al 7075 and Al 1100. For Al 7075, the ratio of 111 to 200 increases from 1.20 to 2.2 close to the crack, implying a change approximately 130%. This suggests that considerable lattice rotation across several grains takes place in front of the crack as a result of fatigue crack growth at room temperature [57].

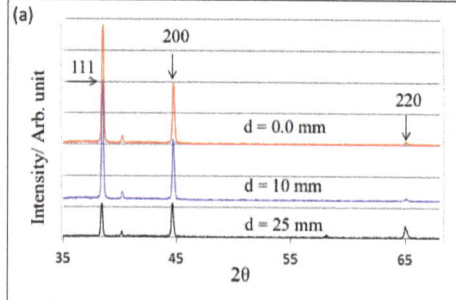 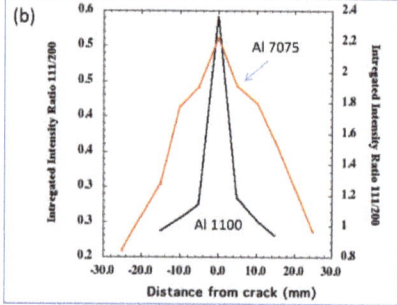

Figure 10. (a) X-ray diffraction (XRD) patterns as a function of distance from the crack showing the relative variations of intensity of 111 and 200 peaks. (b) The integrated intensity ratio of 111 to 200 as a function of position from crack for Al 1100 and Al 7075. The y-axis values for Al 7075 are shown on the right side.

The compressive residual stress enhances the fatigue lifetime [61], however, the tensile residual stress is highly damaging under fatigue loading [62]. Thus, the residual stress can affect the crack growth by influencing the stress intensity factor, the mean stress and the fatigue life [63,64]. In addition, the local residual stress could play a significant role than the overall large scale distribution of the residual stresses. To estimate the residual stress, the elastic strain has been experimentally obtained with the XRD technique around the crack. The residual stresses were obtained from the slope and the intercept of the d vs. $sin^2(\psi)$ lines [59] as a function of distance from the crack. The slope has been observed to be positive at a number of locations for Al 7075, indicating that the stress is tensile in most places, and the residual stresses increase from 60 MPa to 195 MPa at 5 mm from the crack, and then decreases to 165 MPa close to the crack for Al 7075 (see Figure 11). The stress measured in front of the crack tip is ≈ 220 MPa. For Al 1100, the stresses decrease from 38 MPa tensile at 25 mm to 17 MPa compressive at 5 mm from the crack. It is below 5 MPa tensile close to the crack (see Figure 11).

In Section 3, we have given a detailed description of analytical progress in developing dislocation crack interaction. Although the very nature these analyses are often two dimensional and highly idealized model of the fatigue process, they provide valuable insight of the interactions, and also mathematical expressions of the various fatigue parameters, which can be experimentally tested. As mentioned before in our opinion the most interesting work during the last few years has been development of discrete modeling approach by Pippan et al. [26,27], Deshpande and co-workers [19], and also continuum modeling by Mastorakos and Zbib [20]. They used a three- dimensional analysis, which is more appropriate and realistic, but it does not make use of discrete dislocations. On the experimental side, as seen in Section 4, there is need to use many sophisticated and, in some cases, new analytical techniques to obtain a more realistic picture of the fatigue process. New direct evidence is presented for the cracks as a major source of dislocations taking part in the fracture and fatigue process. From our model, we obtain the size of the plastic zone and an estimate of the critical threshold for further dislocation emission.

A significant lattice rotation has been observed in the plastic zone of a fatigue crack. We ascribe such rotation to glide of large number of dislocations in front of crack. As we approach the crack, the residual stress increases gradually around the fatigue crack by ≈ 200% for Al-7075, and decreases by ≈ 80% for Al-1100. Such change in residual stress cannot be explained by the difference in dislocation density alone. We demonstrate that the deformation associated with the lattice rotation is a major factor controlling the residual stress [59].

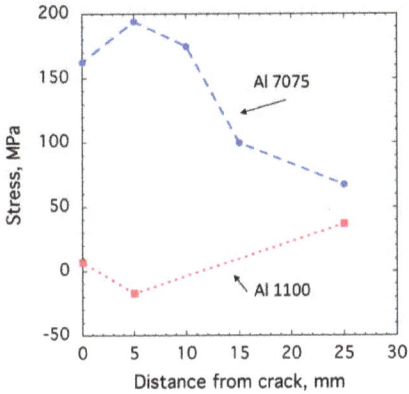

Figure 11. The estimated residual stress as a function of position from crack.

This review, we hope, establishes the fact that cracks are the primary source of dislocations. In fact, in our own work no secondary source of dislocation ahead of the crack was found. On the theoretical side, it appears that analytical modeling of the dislocation-crack interaction is much more complex than usually assumed. Firstly, there are more than one dislocation source at the crack tips. Secondly, the 3D nature of the process is hard to model and thirdly, the dislocations are curved. The superdislocation model introduced by Lin and Thompson may simplify some of the analytic difficulties. A relatively new result, that is, the rotation of the grains ahead of the crack, even when the grains are not in the nano range (≤10 nm) was shown to be an important part of the fatigue process, and should be taken into consideration in making fatigue predictions.

6. Concluding Remarks

The detailed picture of the deformation and fatigue processes has now been well described in the literature. A semi-emperical understanding of various features involved in these processes is also available. Using empirical techniques and experimental methods, fairly accurate predictions can sometimes be made [14]. A macro picture of the process has also emerged, where dislocation density and dislocation configurations play a significant part. For an excellent, but brief recent summary, see the review by Mughrabi [65]. For most of the purposes the above described knowledge of the phenomenon is sufficient.

Despite the progress so far made, our basic understanding of many features of the processes is lacking. In the absence of the basic understanding, dislocation modeling has so far not become a predictive tool. It has been mostly used in many cases to provide quantitative understanding of the experiments or give some rule of thumb for practical applications. We want therefore to strive a more basic understanding of the processes involved. In other words, we want to study the issues involved at a more fundamental level using modern analytical techniques and to use as far as possible rigorous mathematical techniques to analyze the dislocation configurations and their role.

Towards this goal, some fundamental studies have been undertaken, which are briefly described in this review. In 2001, Riemelmoser et al. [28] gave an overview of the dislocations modeling of cracks and in 2003 Pipan et al. [66] published a chapter in much greater detail on dislocation models of fatigue crack growth. Another review dealing with more applied aspects of the basic theoretical models, published in 2011 by Bhat and Patibandla [67] is also very useful. A short review of 2D modeling, which is most common, is provided by Olarnrithinun, in 2013 [68].

One aspect of the modeling not mentioned in our review is the atomistic modeling of the fatigue process. This technique may be very useful in the future as the capacity of the computing process increases. A review of this technique applied to nanostructurally small cracks is available by Horstemeyer et al. [69]. In summary, much remains to be discovered and investigated before a comprehensive methodology for predicting fatigue from first principles can be made available.

Funding: This research was funded by the Office of Naval Research (ONR) through 6.1 program at Naval Research Laboratory (NRL), Washington DC 20375.

Acknowledgments: We also thank Ronald Armstrong for critical comments and encouragement in the preparation of this paper.

Conflicts of Interest: The authors declare no conflict of interest.

References

1. Findlay, S.J.; Harrison, N.D. Why aircraft fail. *Mater. Today* **2002**, *55*, 18–25. [CrossRef]
2. Basinski, Z.S.; Basinski, S.J. Low amplitude fatigue of copper single crystals-III PSB sections. *Acta Metall.* **1985**, *33*, 1319–1327. [CrossRef]
3. Mott, N.F. A theory of the origin of fatigue cracks. *Acta Metall.* **1958**, *6*, 195–197. [CrossRef]
4. Antonopoulos, J.G.; Brown, L.M.; Winter, A.T. Vacancy dipoles in fatigued copper. *Philos. Mag.* **1976**, *34*, 549–563. [CrossRef]
5. Differt, K.U.; Essmann, U.; Mughrabi, H. A model of extrusions and intrusions in fatigued metals-II surface roughening by random irreversible slip. *Philos. Mag. A* **1986**, *54*, 237–258. [CrossRef]
6. Neumann, P. Coarse slip model of fatigue. *Acta Metall.* **1969**, *17*, 1219–1225 [CrossRef]
7. Kobayashi, S.; Ohr, S.M. *Proceedings of the 37th Annual Meeting Electron Microscopy Society of America*; Bailey, G.W., Ed.; Claitors Publishing: Baton Rouge, LA, USA, 1979; pp. 424–425.
8. Kobayashi, S.; Ohr, S.M. In-situ Fracture Experiments in BCC Metals. *Philos. Mag.* **1980**, *A42*, 763–772. [CrossRef]
9. Kobayashi, S.; Ohr, S.M. In-situ Observations of the Formation of Plastic Zone Ahead of a Crack Tip in Copper. *Scr. Metall.* **1981**, *15*, 343–348. [CrossRef]
10. Park, C.G.; Lee, S.; Chang, Y.W. Mechanical behavior of materials. In Proceedings of the 6th International Conference, Kyoto, Japan, 29 July–2 August 1991; Volume 4, pp. 3–9.
11. Goswami, R.; Pande, C.S. Investigations of Crack-Dislocation Interactions Ahead of Mode-III Crack. *Mater. Sci. Eng. A* **2015**, *627*, 217–222. [CrossRef]
12. Du, K.L.; Lu, J.B.; Li, R.W. A continuous dislocation model of mode-I crack. In Proceedings of the 15th ASCE Engineering Mechanics Conference, New York, NY, USA, 2–5 June 2002; Columbia University: New York, NY, USA, 2002; pp. 1–8.
13. Chang, S.J.; Ohr, S.M. Dislocation free zone model of fracture. *J. Appl. Phys.* **1981**, *52*, 7174–7181. [CrossRef]
14. Suresh, S. *Fatigue of Materials*, 2nd ed.; Cambridge University Press: Cambridge, UK, 1998.
15. Pippan, R. Dislocation emission and fatigue crack growth threshold. *Acta Metall. Mater.* **1991**, *39*, 255–262. [CrossRef]
16. Pippan, R. The condition for the cyclic plastic deformation of the crack tip: The influence of dislocation obstacles. *Int. J. Fract.* **1992**, *58*, 305–318. [CrossRef]
17. Riemelmoser, F.O.; Pippan, R.; Stuwe, H.P. A comparison of a discrete dislocation model and a continuous description of cyclic crack tip plasticity. *Int. J. Fract.* **1997**, *85*, 157–168. [CrossRef]
18. Riemelmoser, F.O.; Pippan, R.; Stuwe, H.P. An argument for a cycle-by-cycle propagation of fatigue cracks at small stress intensity ranges. *Acta Mater.* **1998**, *46*, 1793–1799. [CrossRef]

19. Deshpande, V.S.; Needleman, A.; Giessen, E.A discrete dislocation analysis of near-threshold fatigue crack growth. *Acta Mater.* **2001**, *49*, 3189–3203. [CrossRef]
20. Mastorakos, I.N.; Zbib, H.M. Dislocation-cracks interaction during fatigue: A discrete dislocation dynamics simulation. *J. Met.* **2008**, *60*, 59–63 [CrossRef]
21. Curtin, W.A.; Deshpande, V.S.; Needleman, A.; Giessen, E.; Wallin, M. Hybrid discrete dislocation models for fatigue crack growth. *Int. J. Fatigue* **2010**, *32*, 1511–1520. [CrossRef]
22. Wilkinson, A.J.; Roberts, S.G. A dislocation model for the two critical stress intensities required for threshold fatigue crack propagation. *Scr. Mater.* **1996**, *35*, 1365–1371. [CrossRef]
23. Wilkinson, A.J.; Roberts, S.G.; Hirsch, P.B. Modelling the threshold conditions for propagation of stage-I fatigue cracks. *Acta Mater.* **1998**, *46*, 379–390. [CrossRef]
24. Doquet, V. Crack initiation mechanisms in torsional fatigue. *Fract. Eng. Mater. Struct.* **1998**, *21*, 661–672. [CrossRef]
25. Giessen, E.; Deshpande, V.S.; Cleveringa, R.P.; Needleman, A. Discrete dislocation plasticity and crack tip fields in single crystals. *J. Mech. Phys. Solids* **2001**, *49*, 2133–2153. [CrossRef]
26. Pippan, R.; Riemelmoser, F.O.; Weinhandl, H.; Kreuzer, H. Plasticity-induced crack closure under plane-strain conditions in the near-threshold regime. *Philos. Mag.* **2002**, *82*, 3299–3309. [CrossRef]
27. Pippan, R.; Riemelmoser, F.O. Dislocation shielding of fatigue cracks. *Metallkunde* **1995**, *86*, 823–826.
28. Riemelmoser, F.O.; Gumbsch, P.; Pippan, R. Dislocation modelling of fatigue cracks: An overview. *Mater. Trans. A* **2001**, *42*, 2–13. [CrossRef]
29. Hansson, P.; Melin, S. Simulation of simplified zigzag crack paths emerging during fatigue crack growth. *Eng. Fract. Mech.* **2008**, *75*, 1400–1411. [CrossRef]
30. Bjerkn, C.; Melin, S. A tool to model short crack fatigue growth using a discrete dislocation formulation. *Int. J. Fatigue* **2003**, *25*, 559–566. [CrossRef]
31. Bjerkn, C.; Melin, S. A study of the influence of grain boundaries on short crack growth during varying load using a dislocation technique. *Eng. Fract. Mech.* **2004**, *71*, 2215–2227. [CrossRef]
32. Hansson, P.; Melin, S. Dislocation-based modelling of the growth of a microstructurally short crack by single shear due to fatigue loading. *Int. J. Fatigue* **2005**, *27*, 347–356. [CrossRef]
33. Groh, S; Olarnrithinum, S.; Curtin, W.A.; Needleman, A.; Deshpande, V.S.; Giessen, E. Fatigue crack growth from a cracked elastic particle into a ductile matrix. *Philos. Mag.* **2008**, *88*, 3565–3583. [CrossRef]
34. Kunkler, B.; Duber, O.; Koster, P.; Krupp, U.; Fritzen, C.P.; Christ, H.J. Modeling of short crack propagation - Transition from stage I to stage II. *Eng. Fract. Mech.* **2008**, *75*, 715–725. [CrossRef]
35. Ohr. S.M.; Narayan, J. Electron Microscope Observation of Shear Cracks in Stainless Steel Single Crystals. *Philos. Mag.* **1980**, *41*, 81–89. [CrossRef]
36. Horton, J.A.; Ohr, S.M. Determination of emission condition from the experiments and theory. *Scr. Metall.* **1982**, *16*, 621–626. [CrossRef]
37. Horton J.A.; Ohr, S.M. TEM observations of dislocation emission at crack tip in aluminum. *J. Mater. Sci.* **1982**, *17*, 3140–3148. [CrossRef]
38. Ohr, S.M. An electron microscope study of crack tip deformation and its impact on the dislocation theory of fracture. *Mater. Sci. Eng.* **1985**, *72*, 1–35. [CrossRef]
39. Majumdar, B.; Burns, S.A. Crack tip shielding-anelastic theory of dislocation and dislo- cation arrays near a sharp crack. *Acta Metall.* **1981**, *29*, 579–588. [CrossRef]
40. Majumdar, B.; Burns, S.A. A Griffith crack shielded by a dislocation pile-up. *Int. J. Fract. Mech.* **1983**, *21*, 229–240. [CrossRef]
41. Dai, S.H.; Li, J.C.M. Dislocation free zone at the crack tip. *Scr. Metall.* **1982**, *16*, 183–188. [CrossRef]
42. Thomson, R. Physics of fracture. *J. Phys. Chem. Solids* **1987**, *48*, 965–983. [CrossRef]
43. Sadananda, K.; Glinka, G. Dislocation processes that affect kinetics of fatigue crack growth. *Philos. Mag.* **2005**, *85*, 189–203. [CrossRef]
44. Bilby, B.; Cottrell, A.; Swinden, K. The spread of plastic yield from a notch. *Proc. Roy. Soc. A* **1963**, *272*, 304–314.
45. Lin, I.H.; Thomson, R. Cleavage, dislocation emission, and shielding for cracks under general loading. *Acta Metall.* **1986**, *34*, 187–206. [CrossRef]
46. Weertman, J. Fracture mechanics-unified view for Griffith-Irwin-Orowan cracks. *Acta Metall.* **1978**, *26*, 1731–1738. [CrossRef]

47. Weertman, J. Fracture stress obtained from the elastic crack tip enclave model. *J. Mater. Sci.* **1980**, *15*, 1306–1310. [CrossRef]
48. Weertman, J.; Lin, I.H.; Thomson, R. Double slip plane crack model. *Acta Metall.* **1983**, *31*, 473–482. [CrossRef]
49. Weertman, J. Dislocation emission into a mode III plastic zone. *Scr. Metall.* **1986**, *20*, 1483–1488. [CrossRef]
50. Masumura, R.A.; Pande, C.S.; Chou, Y.T. Model for interaction of dislocation arrays with a crack. *Int. J. Fatigue* **2005**, *27*, 1170–1174. [CrossRef]
51. Pande, C.S.; Masumura, R.A.; Chou, Y.T. Shielding of crack tips by inclined pile-ups of dislocations. *Acta Metall.* **1988**, *36*, 49–54. [CrossRef]
52. Muskhelishvili, N. *Singular Integral Equations*; P. Noordhoof Ltd.: Groningen, The Netherlands, 1953.
53. Zhang, T.Y.; Tong, P.; Ouyang, H.; Lee, S. Interaction of an edge dislocation with a wedge crack. *J. Appl. Phys.* **1995**, *78*, 4873–4880. [CrossRef]
54. Pande, C.S.; Masumura, R.A.; Chou, Y.T. Shielding of crack tips by critical parameters for fatigue damage. *Int. J. Fatigue* **2001**, *23*, 1170–1174.
55. Vasudevan, A.K.; Sadananda, K.; Glinka, G. Critical parameters for fatigue damage. *Int. J. Fatigue* **2001**, *23*, S39–S53. [CrossRef]
56. Rice, J.R.; Thomson, R. Ductile versus brittle behavior of crystals. *Philos. Mag.* **1974**, *29*, 73–97 [CrossRef]
57. Goswami, R.; Qadri, S.B. ; Pande, C.S. Fatigue mediated lattice rotation in Al Alloys. *Acta Mater.* **2017**, *129*, 33–40. [CrossRef]
58. Goswami, R.; Feng, C.R.; Qadri, S.B.; Pande, C.S. Fatigue-assisted grain growth in Al alloys. *Sci. Rep.* **2017**, *7*, 10179–10186. [CrossRef] [PubMed]
59. Goswami, R.; Qadri, S.B.; Pande, C.S. Residual Stresses and Localized Lattice Rotation Under Fatigue Loading. *Mater. Sci. Eng. A* **2019**, *763*, 113–138. [CrossRef]
60. Lam, Y.C.; Lian, K.S. Effect of residual stress and its redistribution on fatigue crack growth. *Theor. Appl. Fract. Mech.* **1989**, *12*, 59–66. [CrossRef]
61. LaRue, J.E.; Daniewicz, S.R. The effect of residual stress on fatigue crack growth rate in AISI 304LN stainless steel. *Int. J. Fatigue* **2007**, *29*, 508–515. [CrossRef]
62. Reid, C.N. A Method of mapping residual stress in a compact tension specimen. *Scr. Metall.* **1988**, *22*, 451–456. [CrossRef]
63. Schindler, H.J. Determination of Residual Stress Distributions from Measured Stress Intensity Factors. *Int. J. Fract.* **995**, *74*, R23–R30. [CrossRef]
64. Schindler, H.J.; Cheng, W.; Finnie, I. Experimental determination of stress intensity factors due to residual stresses. *Exp. Mech.* **1997**, *37*, 272–277. [CrossRef]
65. Mughrabi, H. Revisiting Steady-State Monotonic and Cyclic Deformation: Emphasizing the Quasi-Stationary State of Deformation. *Metall. Mater. Trans. A* **2020**, *51A*, 1441–1456. [CrossRef]
66. Pippan, R.; Riemelmoser, F.O. Modeling of Fatigue Crack Growth: Dislocation Models. *Compr. Struct. Integr.* **2003**, *191*–207. [CrossRef]
67. Bhat, S.; Patibandla, R. *Metal Fatigue and Basic Theoretical Models: A Review, Alloy Steel-Properties and Use*; Morales, E.V., Ed.; InTech: Lagos, Nigeria, 2011.
68. Olarnrithinun, S. A Short Review of 2D-Discrete Dislocation Modeling for Fracture/Fatigue. *AIJSTPME* **2013**, *6*, 45–57.
69. Horstemeyer, M.F.; Farkas, D.; Kim, S.; Tang, T.; Potirniche, G. Nanostructurally small cracks (NSC): A review on atomistic modeling of fatigue. *Int. J. Fatigue* **2010**, *32*, 1473–1502. [CrossRef]

© 2020 by the authors. Licensee MDPI, Basel, Switzerland. This article is an open access article distributed under the terms and conditions of the Creative Commons Attribution (CC BY) license (http://creativecommons.org/licenses/by/4.0/).

Creative

Holistic Approach on the Research of Yielding, Creep and Fatigue Crack Growth Rate of Metals Based on Simplified Model of Dislocation Group Dynamics

A. Toshimitsu Yokobori, Jr.

Strategic Innovation and Research Center, Teikyo University, 2-11-1 Kaga Itabashi-ku, Tokyo 173-8605, Japan; toshi.yokobori@med.teikyo-u.ac.jp

Received: 5 May 2020; Accepted: 2 July 2020; Published: 3 August 2020

Abstract: The simplified model of numerical analyses of discrete dislocation motion and emission from a stressed source was applied to predict the yield stress, dislocation creep, and fatigue crack growth rate of metals dominated by dislocation motion. The results obtained by these numerical analyses enabled us to link various dynamical effects on the yield stress, dislocation creep, and fatigue crack growth rate with the experimental results of macroscopic phenomena, as well as to link them with theoretical results obtained by the concept of static, continuously distributed infinitesimal dislocations for the equilibrium state under low strain or stress rate conditions. This will be useful to holistic research approaches with concern for time and space scales, that is, in a time scale ranging from results under high strain rate condition to those under static or low strain rate condition, and in a space scale ranging from meso-scale to macro-scale mechanics. The originality of results obtained by these analyses were found by deriving the analytical formulations of number of dislocation emitted from a stressed source and a local dynamic stress intensity factor at the pile-up site of dislocations as a function of applied stress or stress rate and temperature material constants. This enabled us to develop the predictive law of yield stress, creep deformation rate, and fatigue crack growth rate of metals dominated by dislocation motion. Especially, yielding phenomena such as the stress rate and grain size dependence of yield stress and the delayed time of yielding were clarified as a holistic phenomenon composed of sequential processes of dislocation release from a solute atom, dislocation group moving, and stress concentration by pile-up at the grain boundary.

Keywords: holistic approach; dislocation group dynamics; dynamic factor; dislocation pile-up; yield stress; dislocation creep; fatigue crack growth rate

1. Introduction

The purpose of the research of dislocation mechanics is considered to have two directionalities, that is, application to the research of materials science and application to the research of the strength of materials such as that of yielding and fatigue crack growth rate.

The former closely relates to micro plasticity, such as the conditions of dislocation emission, annihilation, and cross slip, and this research was developed in the manner of the modern dislocation dynamics [1].

The latter closely relates to connect problems of the strength and fracture of materials with the macro scale [2–5].

This article is related to the fracture of materials. For this case, the numerical results of the number of moving dislocations emitted from a stressed source and local stress concentration caused by a dislocation pile-up were necessary to be formulated as an analytical function of applied stress or stress rate, temperature, and material constants [2,3].

Since the behaviors of dislocation group motion have a scale of μm at the meso-scale level, that is, the intermediate scale of nm (an individual dislocation) and mm (crack length) scales, which is a comparable scale with grain size. At this scale, the simplification of the model of analysis is considered to be more convenient to derive a predictive law of the strength and fracture of metals.

To fulfil this purpose, in our research, the establishment of a predictive theory of strength and fracture of materials was conducted by conducting the simplification of the model of analysis [3] and verification with experimental results [3,6].

Yielding phenomena such as stress rate and grain size dependence of yield stress and delayed time of yielding were especially clarified as a holistic phenomenon composed of sequential processes of dislocation release from a solute atom [7,8], dislocation group moving [6], and stress concentration by a pile-up at the grain boundary [2–4].

2. Dislocation Groups Dynamics Aimed for Applications to Problems of Yielding, Creep, and Fatigue [2,9,10]

2.1. Model, Basic Equation and Method of Analysis [2,9,10]

Plastic deformation is caused by dislocation group motion emitted from a stressed source.

It closely relates to plastic yielding and fatigue crack growth dominated by discrete dislocations emitted from a stressed source near a crack tip. Pioneering works on the analysis of discrete dislocation group dynamics emitted from a stressed source have been conducted [4,11,12], but a power law equation has been adopted between the isolated dislocation velocity and the stress for practical application to the strength of materials such as the yield stress, creep rate, and fatigue crack growth rate; this is given by Equation (1), which is related to experimental equations [4].

Equations of discrete moving dislocation groups are given by Equations (1) and (2).

$$V_i = \frac{dx_i}{dt} = M\tau_{eff,i}{}^m \tag{1}$$

$$\tau_{eff,i} = \dot{\tau}t + A \sum_{\substack{j=1 \\ i \neq j}}^{n} \frac{1}{x_i - x_j} \quad (i = 1 \sim n) \tag{2}$$

In the equations, V_i and x_i are velocity and position of individual dislocations in a linear array, respectively, and M and m are the material constants of an isolated dislocation given by the experimental Equation (3) [13]. The calculation starts at the time of $t = 0.0$ s.

$$v = v_0 \left(\frac{\tau}{\tau_0^*}\right)^m V \tag{3}$$

In Equation (3), $\dot{\tau}$ is the stress rate; t is the time of stress application; $\tau_{eff,i}$ is the effective stress exerted on individual dislocations in terms of shear stress; and $A = \frac{Gb}{2\pi(1-\nu)}$, where G is the shear modulus, b is the Burgers vector, ν is the Poisson's ratio, $V_0 = 1$ cm/s, and τ_0^* is a constant representing the stress required to give a dislocation velocity $v = 1$ cm/s (resistant stress against the dislocation motion).

A free expansion model of linear dislocation motion emitted from a stressed source, S, is shown in Figure 1 [9,10].

The numerical analyses were conducted as follows [9].

When the effective stress exerted on dislocation source ($x = 0.0$) takes the source activation stress, the new dislocation is originated at $x = 0.0$, and these processes are iterated. Equations (2) and (1) were solved by the Runge–Kutta Merson method. The effective stress exerted on a dislocation source is given by Equation (4).

$$\tau = \dot{\tau}t$$

$$\tau_{eff,s} = \dot{\tau}t - A\sum_{j=1}^{n}\frac{1}{x_j} \tag{4}$$

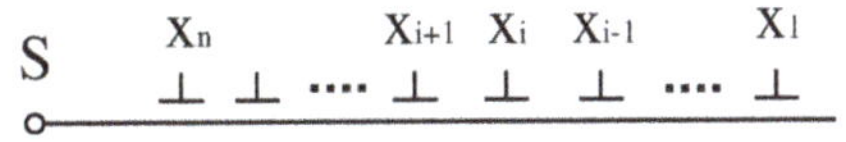

```
S    Xn           Xi+1  Xi  Xi-1        X1
o—  ⊥ ⊥  ....  ⊥   ⊥   ⊥   ....  ⊥
```

Figure 1. Free expansion model of linear dislocation motion emitted from a stressed source, S.

2.2. Discrete Dislocation Groups Dynamics of Free Expansion and Similarity Law of Dislocation Flow [9,10]

From the analysis, the ratio of positions, velocity and effective stress of individual dislocation in the dislocation array to those of an isolated dislocation, such as $\frac{x_i}{x_{iso}}$, $\frac{v_i}{v_{iso}}$, and $\frac{\tau_{effi}}{\tau_{iso}}$, were found to be dominated by $\Theta = \left(\frac{\dot{\tau}}{\tau_0}\right)^{\frac{(m+1)}{(m+2)}}\theta$, which is named the dynamic factor [9,10]. Here, τ is the applied stress acting on dislocations in the dislocation array, θ is the non-dimensional time controlled by t_0, and t_0 is the time of an isolated dislocation moving the distance, l. It is given by $t_0 = \left[\frac{(m+1)l}{M\tau_0^m}\right]^{\frac{1}{(m+1)}}$ [2,9,10] by using Equation (3). In this analysis, l was taken as the length of 0.01 mm.

Furthermore, the number of dislocations emitted from a stressed source were also found to be given by the dynamic factor, Θ, which is a non-dimensional character, as shown in Equation (5a); the dimensional parameter η is shown in Equation (5b) [9,10].

$$N = A\Theta^{m+1} \tag{5a}$$

$$N = A_0\eta^{m+1} \tag{5b}$$

$$\eta = \dot{\tau}^{\frac{m+1}{m+2}}t = \tau\dot{\tau}^{-\frac{1}{m+2}}$$

In the equation, A and A_0 are non-dimensional and dimensional constants, respectively. By dimensional analysis, N is given by Equation (6a) [9,10].

$$N = \gamma(m)\left(\frac{M}{Gb}\right)^{\left(\frac{m+1}{m+2}\right)}\eta^{m+1} \tag{6a}$$

$$\gamma(m) = 1.4m^{-1.45} \tag{6b}$$

In these equations, $\gamma(m)$ is a non-dimensional function depending on m.

The velocity of an isolated dislocation is given by thermally activated process, as shown in Equation (2) [14].

$$v = A_1\exp\left(-\frac{H}{RT}\right) \tag{7}$$

$$H = H_k\left(1 + \frac{1}{4}\ln\frac{16\tau_p^0}{\pi\tau}\right) \tag{8}$$

In these equations, A_1 is a constant, H_k is the kink energy, τ is the applied stress, τ_p^0 is the Peierls stress at 0° K, and T is the absolute temperature. By substituting Equation (8) into Equation (7) and comparing Equation (7) with Equation (9), which is the experimental equation of an isolated dislocation, Equations (10a) and (10d) could be obtained.

$$v = M\tau^m \tag{9}$$

$$m = \frac{H_k}{4kT} \tag{10a}$$

$$M = v_0 \left(\frac{1}{\tau_0^*}\right)^m \tag{10b}$$

$$\tau_0^* = \tau_{00}\left(\frac{A_1}{v_0}\right)^{-1/m} \tag{10c}$$

$$\tau_{00} = e^4\left(\frac{16}{\pi}\right)\tau_p^0 \tag{10d}$$

Using Equations (10a)–(10d), Equation (6a) was able to be rewritten as Equation (11). From this equation, the number of dislocations emitted from a stressed source was found to be dominated by a thermally activated process [10].

$$N = A_* t_a^{\frac{m;1}{m+2}} \left(\frac{\tau}{G}\right)^{\frac{m;1}{m+2}} \exp\left\{-\frac{m+1}{m+2} H_k \ln\left(\frac{\tau_{00}}{\tau}\right)/4kT\right\} \tag{11}$$

$$\text{where} \quad A_* = \gamma(m)\left(\frac{b}{A_1}\right)^{-\frac{m+1}{m+2}} \tag{12}$$

2.3. Dislocation Pile-Up Induced by Local Stress Field [2]

Some previous research has treated analyses of dislocation pile-up [4,11,15], but there has not been so much research that has considered the application to fracture mechanics description.

In this section, numerical analyses were conducted on the dynamic piling-up of discrete dislocations emitted from a stressed source and on the dynamic stress intensity factor caused by discrete moving dislocations in a pile-up.

2.3.1. Model, Basic Equation and Analysis [2]

Until the lead dislocation in the array arrives at a barrier, such as grain boundary, dislocations will emit from a stressed source and move freely except for the interactions between dislocations, as shown in Figure 1. Equations of the motion of dislocation groups are given by Equations (1) and (2). After the arrival of the lead dislocation at the barrier, it is locked and the trailing dislocations pile-up against the barrier, as shown in Figure 2.

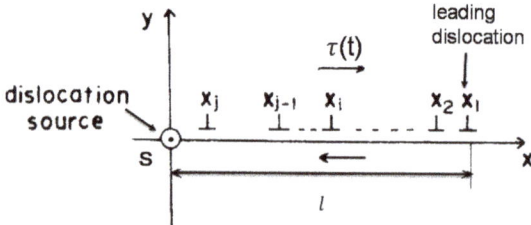

Figure 2. Discrete dislocation pile-up model with emission from a stressed source.

Equations of effective stress exerted on each dislocation in the array for the case of dislocation group pile-up are given by Equation (13).

$$\tau_{eff,i} = \tau t + A\left(\frac{1}{x_i - l} + \sum_{\substack{j=2 \\ i \neq j}}^{n} \frac{1}{x_i - x_j}\right) \quad i = 1 \sim n \tag{13}$$

The equation of motion of each dislocation in the array is calculated by Equation (1).
The effective stress exerted on a dislocation source before and after the leading dislocation arrives at the site of pile-up is given by Equations (14a) and (14b), respectively.

$$\tau_{eff,s} = \dot{\tau}t - A \sum_{j=2}^{n} \frac{1}{x_j}, \ (x_1 < l) \tag{14a}$$

$$\tau_{eff,s} = \dot{\tau}t - A(\frac{1}{l} + \sum_{j=2}^{\dot{n}} \frac{1}{x_j}), \ (x_1 = l) \tag{14b}$$

The stress distribution, $\tau(x,t)$, caused by dynamical piling up in the region of $x > l$ is shown in Figure 3 and is given by Equation (15).

$$\tau(x,t) = \frac{A}{x-l} + A\sum_{i=2}^{n} \frac{1}{x-x_i} + \dot{\tau}t \tag{15}$$

The dynamic stress intensity factor caused by dislocation pile-up formation is given by Equations (2)–(16).

$$k(t) \cong \sqrt{2\pi(x-l)} \tau(x,t)_{l<x<l(1+\varepsilon)} \tag{16}$$

where $l\varepsilon$ is the small distance in which the stress distribution has the characteristic of

$$\frac{1}{\sqrt{x^*}} (x^* = x - l = l\varepsilon)$$

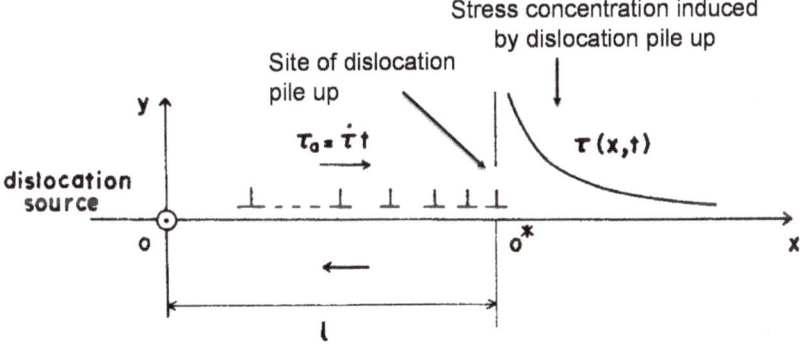

Figure 3. The stress distribution, $\tau(x,t)$, caused by dynamical piling up [3].

2.3.2. Results [2,3]

The numerical results of stress distribution on the slip line $x > l$ near the site of pile-up (barrier) were obtained by Equation (15), as shown in Figures 4 and 5, where $s^* = s - 1$ is the non-dimensional distance from the site of dislocation pile-up (O^* in Figure 3), S is the non-dimensional value of x controlled by l.

When the number of dislocations emitted is as small as shown in Figure 4, the stress distribution near the barrier shows a $1/s^*$ singularity, and with an increase in s^*, a $\frac{1}{\sqrt{s^*}}$ singularity appears but is restricted within a narrow region. ($3 \times 10^{-3} < s^* < 2 \times 10^{-2}$) The characteristics of $\frac{1}{\sqrt{s^*}}$ take minor portion in the stress distribution [2].

On the other hand, when the number of emitted dislocations increases, as shown in Figure 5 (sixty emitted dislocations), the stress distribution shows a singularity of $\frac{1}{\sqrt{S^*}}$, which appears from the vicinity of the barrier; this characteristic region extends up to 5% of the length of slip line [2].

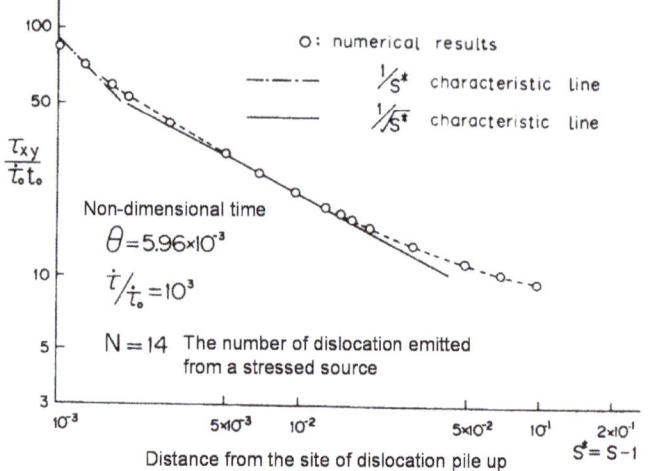

Figure 4. The numerical results of stress distribution on the slip line, x > l (S*), near the site of pile-up (barrier) obtained by Equation (15) for the case of N = 14 (Small number of dislocations) [2].

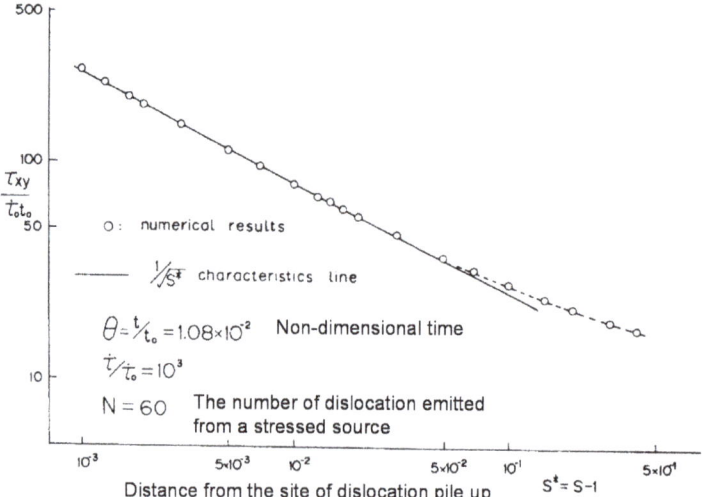

Figure 5. The numerical results of stress distribution on the slip line, x > l (S*), near the site of pile-up (barrier) obtained by Equation (15) for the case of N = 60 (large number of dislocations) [2].

The numerical results of the dynamic stress intensity factor $\overline{K_d}(\theta)$ due to pile-up by discrete dislocation groups dynamics with emission were obtained, as shown in Figure 6, by using Equations (15) and (16) and by the $\frac{1}{\sqrt{S^*}}$ singularity of the stress distribution, as in Figure 5. A static solution, $\overline{K_s}(\theta)$, obtained by the concept of continuously distributed infinitesimal dislocations for the equilibrium pile-up is given by Equation (17) and is also shown in Figure 6 for comparison with the dynamic stress intensity factor, $\overline{K_d}(\theta)$. $\overline{K_s}(\theta)$ and the linear part of $\overline{K_d}(\theta)$ in Figure 6 are written as follows.

$$\overline{K_s}(\theta) = 10^3 \theta \tag{17}$$

$$\overline{K_d}(\theta) = 1.2 \times 10^3 \theta - 5.2 \tag{18}$$

$$\text{where } \theta = \frac{t}{t_0}$$

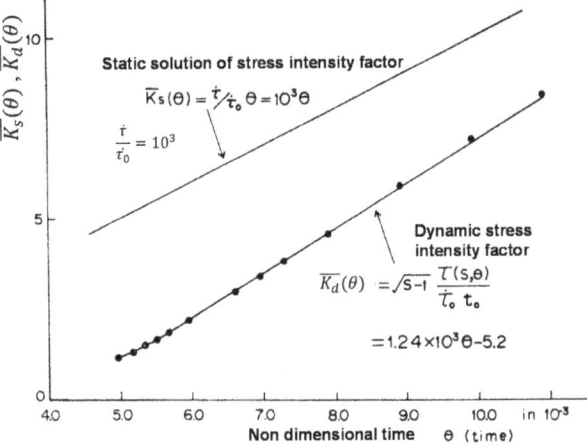

Figure 6. Numerical results of the dynamic stress intensity factor, $\overline{K_d}(\theta)$, due to pile-up by dislocation group dynamics with emission, as well as a comparison with that of a static solution [2].

The values of $\overline{K_d}(\theta)/\overline{K_s}(\theta)$ were plotted against non-dimensional time, θ, as shown in Figure 7. From these results, it can be seen that the dynamic stress intensity factor is smaller than the static one and asymptotically approaches the static one as the number of emitted dislocations increases.

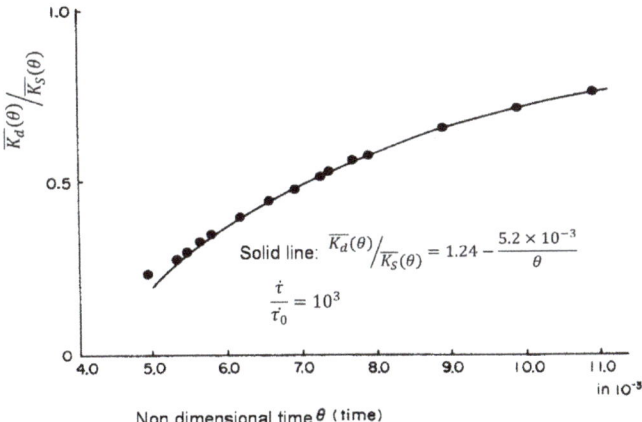

Figure 7. The values of $\overline{K_d}(\theta)/\overline{K_s}(\theta)$ plotted against non-dimensional time, θ [2].

The calculated dynamic dislocation density distribution f_d is shown in Figures 8–11 [3]. On the other hand, the static dislocation density distribution f_s and the number of dislocations N under equilibrium state without emitting are given by the continuous distributed infinitesimal dislocations concept, assuming $f_s(s) = 0$ at the dislocation source and $f_s(s) \to \infty$ at the site of pile-up, respectively. They are given by Equations (19) and (20).

$$f_s(s) = \frac{\tau_a}{\pi A^*}\sqrt{\frac{s}{1-s}} \tag{19}$$

$$N = \int_0^d f_s(s)ds = \frac{\tau_a d}{2A^*} \tag{20}$$

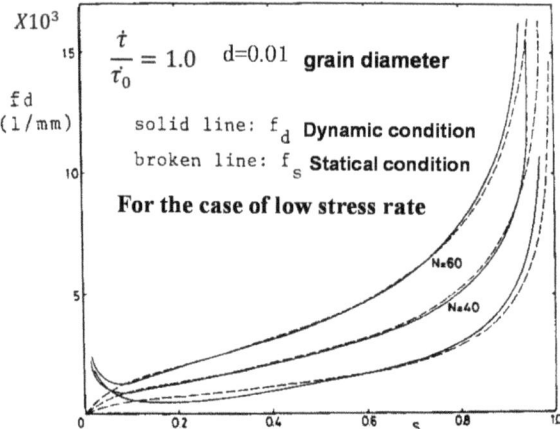

Figure 8. The dynamic dislocation density distribution f_d for grain size (d) = 0.01; non-dimensional stress rate $\left(\frac{\dot{\tau}}{\tau_0}\right) = 10$ [3].

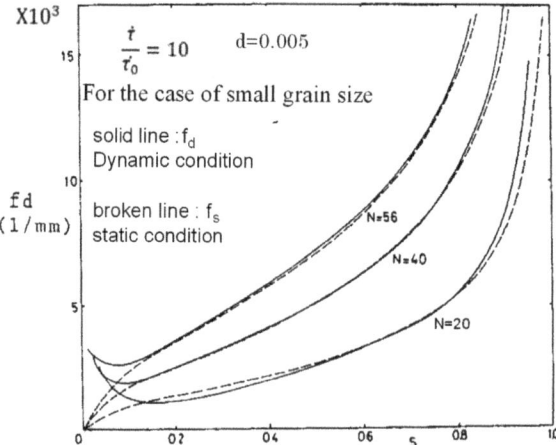

Figure 9. The dynamic dislocation density distribution f_d for d = 0.005; $\frac{\dot{\tau}}{\tau_0} = 10$ [3].

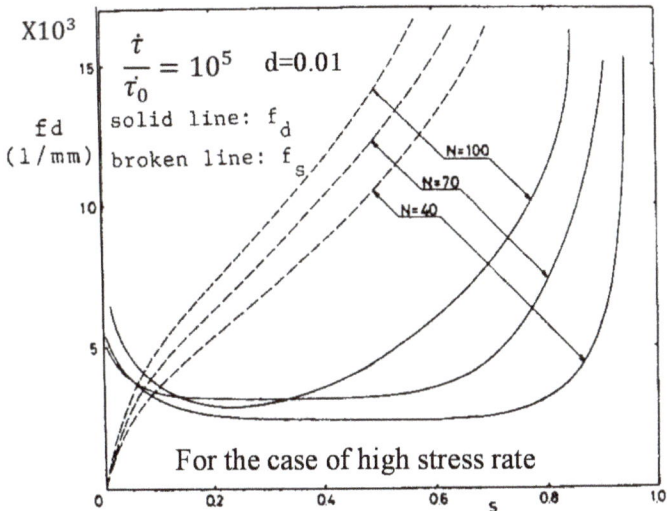

Figure 10. The dynamic dislocation density distribution f_d for d = 0.01; $\frac{\dot{\tau}}{\tau_0} = 10^5$ [3].

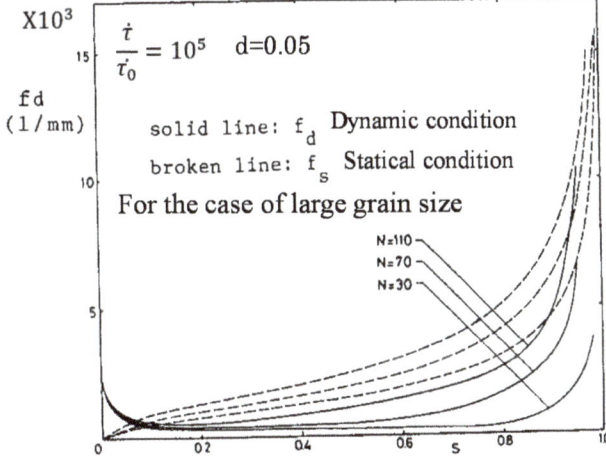

Figure 11. The dynamic dislocation density distribution f_d for d = 0.05; $\frac{\dot{\tau}}{\tau_0} = 10$ [3].

In Figures 8–11, f_s is shown by the dotted line. These results showed when the stress application rate, $\dot{\tau}$, or grain size, d, is small, and a static equilibrium solution based on a continuous distribution of infinitesimal dislocations gives a good approximation, as shown in Figures 8 and 9. However, with increase in $\dot{\tau}$ or d, the dynamic effect becomes more remarkable and f_d becomes smaller than f_s [3], as shown in Figures 10 and 11 [3].

The dynamic stress intensity factor $\overline{K_d}(\theta)$ at the site of pile-up such as grain boundary in non-dimensional form can be obtained using Equation (16). In Figure 12, $\overline{K_d}(\theta)/\overline{K_s}(\theta)$ is plotted against the non-dimensional dynamic factor $\Theta = \left(\frac{\dot{\tau}}{\tau_0}\right)^{\frac{(m+1)}{(m+2)}} \theta$ [9,10]. For the case of iron (m = 3 [13]), the following equation was obtained [3].

$$\frac{K_d}{K_s} = 1.0 - A\exp(-B\Theta) \tag{21}$$

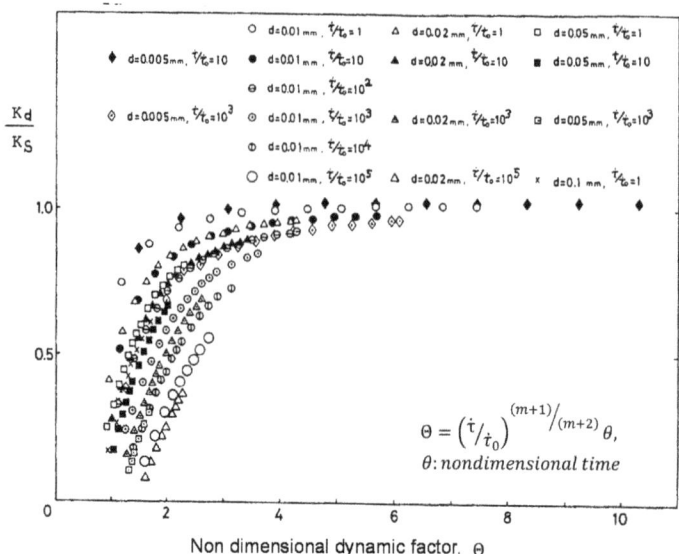

Figure 12. The relationship between dynamic stress intensity factor controlled by static stress intensity factor and the non-dimensional dynamic factor Θ [3].

By using the non-linear least square method, A and B in Equation (21) were obtained as a function of $\frac{\dot{\tau}}{\tau_0}$ and d, respectively, as follows [3].

$$A = \varphi\left(\frac{\dot{\tau}}{\tau_0}, d\right) = 2.35 - 0.0214\left(\frac{\dot{\tau}}{\tau_0}\right)^{-0.172} d^{-0.896} \tag{22}$$

$$B = \psi\left(\frac{\dot{\tau}}{\tau_0}, d\right) = 1.08\left(\frac{\dot{\tau}}{\tau_0}\right)^{-0.0557} \tag{23}$$

The results of Figures 13 and 14 show that Equations (22) and (23) well-represent numerical results.

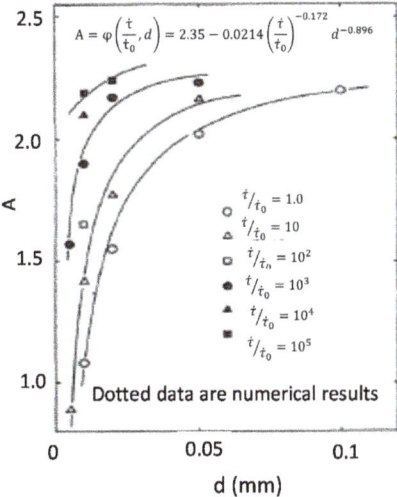

Figure 13. The relationship between a non-dimensional constant (A) and d [3].

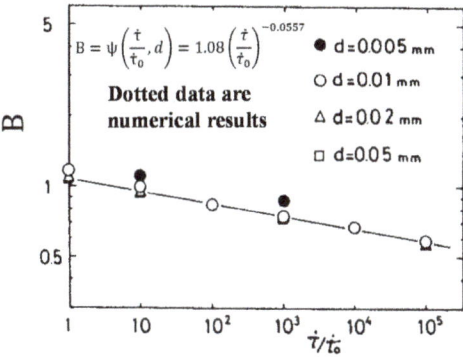

Figure 14. The relationship between B and $\dot{\tau}/\tau_0$ [3].

By substituting Equations (22) and (23) into Equation (21), Equation (24) can be obtained [3].

$$\frac{K_d}{K_s} = 1.0 - \left\{ 2.35 - 0.0214 \left(\frac{\dot{\tau}}{\tau_0}\right)^{-0.172} d^{-0.896} \right\} \exp\left\{ -1.08 \left(\frac{\dot{\tau}}{\tau_0}\right)^{0.75} \frac{t}{t_0} \right\} \quad (24)$$

Figure 15 shows that the calculated values from Equation (24) were found to be in good agreement with data obtained by numerical analyses.

By using dimensional analysis and determining the coefficient of the constant term using the number of material constants, Equation (24) leads to Equation (25) [3].

$$K_d(t) = \sqrt{2\pi d}(\dot{\tau}t) \left[1.0 - \left\{ 2.35 - 0.230 \left(\frac{G}{\dot{\tau}}\right)^{\frac{m}{(m+2)}} \left(\frac{GV_0}{\dot{\tau}b}\right)^{\frac{1}{(m+2)}} \left(\frac{b}{d}\right)^{\frac{(m+1)}{(m+2)}} \right\} \\ \times \exp\left\{ -1.08 \left(\frac{\dot{\tau}t}{\tau_0}\right)^{\frac{m}{(m+2)}} \left(\frac{V_0 t}{(m+1)d}\right)^{\frac{1}{(m+1)}} \right\} \right] \quad (25)$$

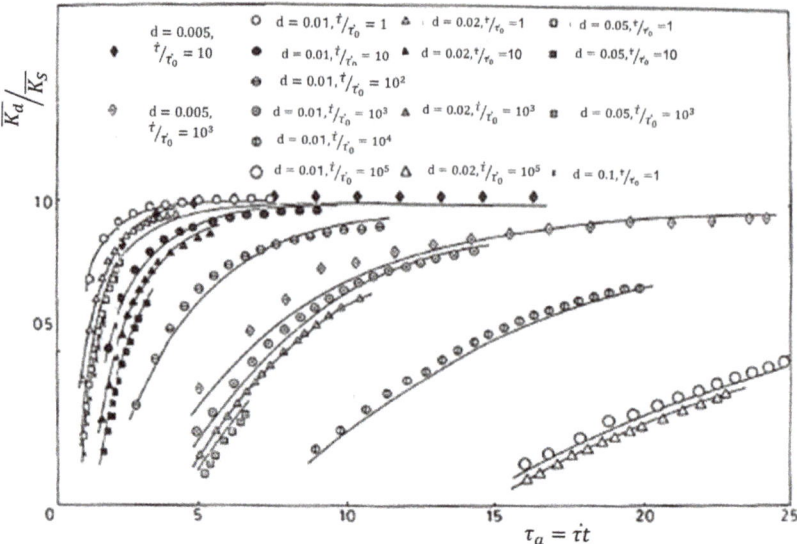

Figure 15. The relationship between dynamic stress intensity factor controlled by static stress intensity and applied stress under constant stress rate condition [3].

2.4. Application to Problem of Yielding

2.4.1. Basic Equations

Cottrell and Bilby described the mechanism of yielding from the view point of dislocation trapping mechanism by solute atoms of carbon [7]; however, the effect of strain rate and temperature have not yet been fully formulated. Concerning ductile fracture of steels, the formulation of the upper yielding point and delay time for yielding are important factors to be analyzed. Takeo Yokobori conducted the formulation of the stress rate dependence of the upper yield point based on a stochastic model analysis that analyzed the releasing process of solute atom from a dislocation [8].

On the other hand, on the basis of the concept that the velocity of an isolated dislocation is proportional to the strain rate of a specimen given by Equation (26), Johnson [15] and Hahn [16] described the yielding phenomenon from the view point of theory of dislocation [15,16]. However, a cleared formulation of yielding phenomenon including the effect of temperature has not yet been conducted. Furthermore, the researchers used the equation of velocity of an isolated dislocation motion by considering that every dislocation moves at the same velocity without interaction between them, as given by Equation (26) [16].

$$\dot{\gamma} = \rho b \bar{v} \tag{26}$$

In the equation, ρ is the dislocation density, b is the Burgers vector, and $\bar{v}$ is the mean velocity of a dislocation in which the equation of velocity of an isolated dislocation was used.

In this section, instead of Equation (26), Equation (27) [4,17], which considers the interaction of dislocations within groups starting with dislocation emission from a stressed source (as calculated by Equations (1) and (2)) was adopted.

$$\dot{\gamma} = \rho b \sum_{i=1}^{n(t)} v_i \tag{27}$$

In Equation (27), $n(t)$ is the number of dislocations emitted from a stressed source at the time of t, i is the dislocate-ion number, and v_i is the velocity of the i th dislocation in the dislocation groups.

By conducting computer simulation using the physical model of Figure 1 and Equations (1)–(4), the summation of non-dimensional velocity of each dislocation in the array was found to be written by Equation (28) and is shown in Figure 16 [17].

$$\sum_{i=1}^{n} \frac{v_i(t)}{v_{iso}} \cong n(t) \tag{28}$$

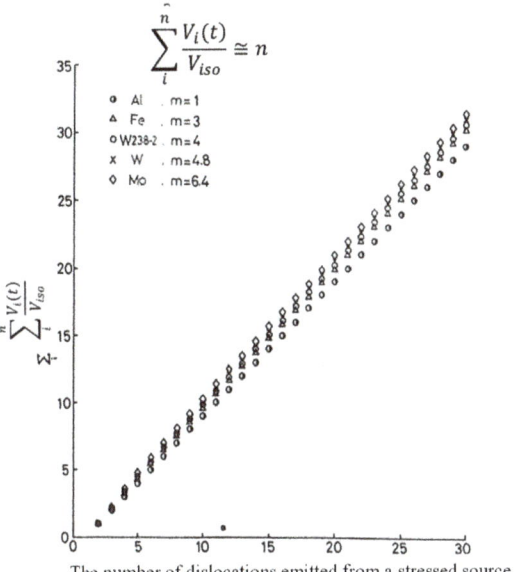

Figure 16. The relationship between $\sum_{i=1}^{n} \frac{v_i(t)}{v_{iso}}$ and the number of dislocation emitted from a stressed source, n [17].

By substituting Equation (28) into Equation (27) and using Equation (29), the plastic strain rate is given by Equation (30) [17].

$$\rho = \rho_0^* \Lambda^* \tag{29}$$

In Equation (29), ρ is the area density dislocation, ρ_0^* is the volume density dislocation, and Λ^* is the average length of dislocation.

$$\dot{\gamma} = \rho_0^* \Lambda^* b n(t) v_0 \left(\frac{\tau}{\tau_0}\right)^m = \overline{\rho}(t) b v \tag{30}$$

In Equation (30), $\overline{\rho}(t) = \rho_0^* \Lambda^* n(t)$, and v is given by Equation (3).

2.4.2. The Application of This Theory to Yielding of Steels

The Delay Time of Yielding

The delay time of yielding under rapid application of constant stress is calculated by Equation (31) as the time of plastic strain, thus taking the specified value [16].

$$\gamma_P = \int_0^t \dot{\gamma} \, dt = const \tag{31}$$

Using Equations (30) and (32), which are the number of dislocations emitted from a stressed source under constant stress condition [6], and by substituting Equations (30) and (32) into Equation (31) and integrating Equation (31), Equation (33) was obtained [17].

$$n(t) = 2.45 m^{-0.865} \left(\frac{b}{v_0 t}\right)^{-\frac{(m+1)}{(m+2)}} \left(\frac{\tau_0^*}{G}\right)^{-\frac{m(m+1)}{(m+2)}} \left(\frac{\tau_a}{G}\right)^{\frac{(m+1)^2}{(m+2)}} \tag{32}$$

where G is the shear modulus. $\tau_a = \tau_Y = \frac{\sigma_Y}{2}$, where σ_Y is the yield stress under uniaxial tensile load, is assumed.

$$t = \left(\frac{\gamma_P}{\rho_0^* \Lambda b^2} \frac{2m+3}{m+2} \frac{m^{0.865}}{2.45}\right)^{\frac{m+2}{2m+3}} \frac{b}{v_0} \left(\frac{G}{\tau_0^*}\right)^{\frac{m+1}{2m+3}} \left(\frac{\sigma_Y}{2\tau_0^*}\right)^{-\frac{(2m^2+4m+1)}{(2m+3)}} \tag{33}$$

By using Equations (7)–(10), Equation (33) was able to be written in the following manner as a function of yield stress, temperature, and material constants [17].

$$t = \left(\frac{0.792 \gamma_P}{\rho_0^* \Lambda b^2}\right)^{0.515} \left(\frac{H_k}{4kT}\right)^{0.446} \frac{b}{A_1} \left(\frac{G}{\tau_{00}}\right)^{0.485} \left(\frac{\sigma_Y}{2\tau_{00}}\right)^{-\left(\frac{H_k}{4kT}\right)} \tag{34}$$

In Equation (34), σ_Y is the yield stress under uniaxial tensile loading. As the criterion of yielding, $\gamma_P = 0.01$ and $\rho_0^* \Lambda^* = 10^8/m^2$ were selected [15], and G = 79.38 GPa, τ_0^* = 169.6 MPa [10], m = 10 [16] were taken as those under room temperature and b = 3×10^{-10}m was selected. Furthermore, for m = 10~30 (for steel) [13], $(2m^2 + 4m + 1)/(2m + 3) \approx m$ was approximately assumed.

A comparison of Equation (34) with experimental data is shown in Figure 17 [17]. Equation (34) was found to well-predict experimental data [18,19]. Furthermore, this equation was in good agreement with that obtained based on dislocation dynamics theory that defined yielding to occur when the dislocation density takes some critical value [6] as follows.

$$t = \left(\frac{\overline{N_0}}{\rho_0^* \Lambda 2.45}\right)^{1.06} \left(\frac{H_k}{4kT}\right)^{0.917} \frac{b}{A_1} \left(\frac{G}{\tau_{00}}\right) \left(\frac{\sigma_Y}{2\tau_{00}}\right)^{-\left(\frac{H_k}{4kT}\right)} \tag{35}$$

In Equation (35), $\overline{N_0}$ is critical dislocation density at the yielding.
This means that the γ_P criterion is identical to the $\overline{N_0}$ criterion.

Furthermore, Equation (34) was found to be in good qualitative agreement with theoretical results [8] based on Cottrell-Bilby's dislocation release mechanism [7] for locking by solute atoms such as carbon or nitrogen, as given by Equation (36). This means that the locking mechanism closely connects with the mechanism of dislocation group dynamics, as described in the following expression.

$$t = t_0 \left(\frac{\sigma_Y}{\sigma_0}\right)^{-\frac{1}{nkT}} \tag{36}$$

where t_0 and σ_0 are material constants.

T. Yokobori found that adopting a friction stress, τ_i, to resist the motion of a dislocation in Equation (36) was very effective in obtaining agreement with experimental results via equations [20]. In this theory, the effect of τ_i on delay time for yielding, as included in Equation (33), was extended to then give Equation (37) [17].

$$t = \left(\frac{\gamma_P}{\rho_0^* \Lambda b^2} \frac{2m+3}{m+2} \frac{m^{0.865}}{2.45}\right)^{\frac{m+2}{2m+3}} \frac{b}{v_0} \left(\frac{G}{\tau_0^* - \tau_i}\right)^{\frac{m+1}{2m+3}} \left(\frac{\sigma_Y}{2(\tau_0^* - \tau_i)}\right)^{-\frac{(2m^2+4m+1)}{(2m+3)}} \tag{37}$$

By adopting $\tau_i = 86.3$ MPa and using Equations (10a)–(10d), Equation (37) was found to well-predict experimental results [19], as shown in Figure 18 [17].

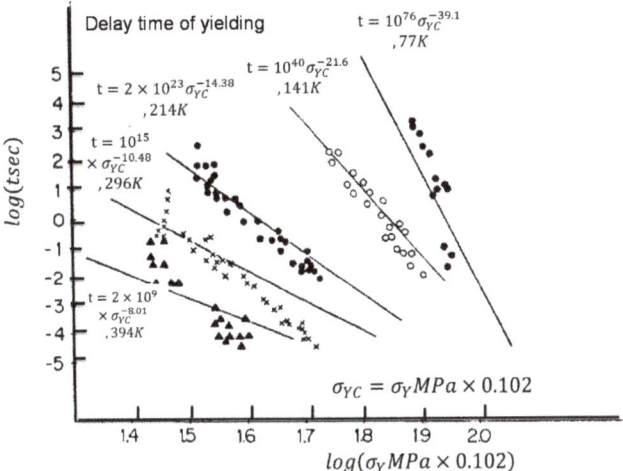

Figure 17. The relationship between the delay time of yielding and yield stress based on dislocation dynamics. Solid lines represent the theoretical results [17]. Dotted lines represent the experimental results [19].

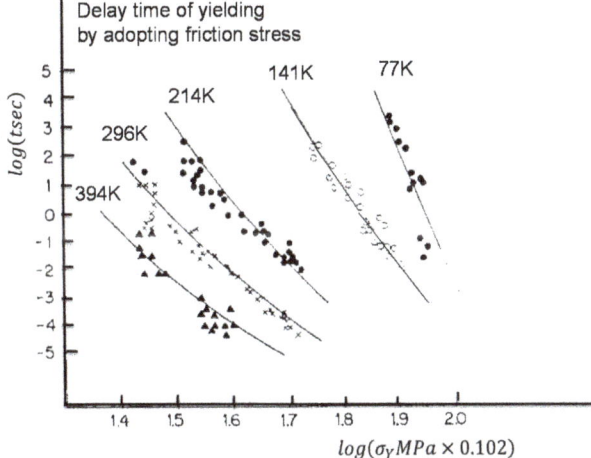

Figure 18. The relationship between the delay time of yielding and yield stress based on dislocation dynamics by accounting for the effect of friction stress of dislocation on delay time. Solid lines represent the theoretical results [17]. Dotted lines represent the experimental results [19].

Furthermore, a previous numerical analysis based on the pile-up behaviors of moving dislocations emitted from a stressed source was conducted [4], and the theoretical relationship between delay time of yielding and yield stress was derived for various grain size [4]. These results were found to be in good agreement with experimental results and were also found to produce similar characteristics to the results given by Figure 18 [17].

From the results mentioned above, the criteria for the various cases of dislocation release from locking by solute atom [8], critical plastic strain [16,17], critical dislocation density [6], and dislocation pile-up at the grain boundary [4] were closely associated with the sequential processes involved in determining the delay time for yielding. Thus, the theoretical results [8] based on Cottrell-Bilby's dislocation release mechanism [7] for dislocation locking by solute atoms (Equation (36)) are considered to be the starting for understanding plastic yielding.

The Applied Stress Rate Dependence of Yield Stress [17]

Strain rate caused by applied stress, $\dot{\gamma}_a$, is given by the summation of plastic strain rate of the specimen, $\dot{\gamma}_P$, and elastic strain rate including those of grips and rigidity of testing machine, $\dot{\gamma}_e$, as follows.

$$\dot{\gamma}_a = \dot{\gamma}_e + \dot{\gamma}_P \tag{38}$$

For the case of a sharp yielding point, $\dot{\gamma}_e \approx 0$ is satisfied at the yield point. Therefore, $\dot{\gamma}_a$ is given by Equation (39).

$$\dot{\gamma}_a = \dot{\gamma}_P \approx A^* \frac{1}{G} \dot{\tau}_a \tag{39}$$

where A^* is a proportional constant. By using Equations (11) and (12) for the number of dislocations emitted from a stressed source under constant stress rate condition and then substituting Equations (26) and (27) into Equation (39), σ_Y can be given by Equation (40) as a function of stress rate and material constants. Furthermore, $\tau_a = \tau_Y = \frac{\sigma_Y}{2}$ and $\dot{\tau} = \frac{\dot{\sigma}}{2}$ were also assumed.

$$\sigma_Y = 2\tau_0^* \left\{ \frac{m^{1.45}}{1.4} \frac{A^*}{\rho_0^* \Lambda b^2} \right\}^{\frac{1}{(2m+1)}} \left(\frac{\tau_0^*}{G} \right)^{-\frac{2(m+1)}{(m+2)(2m+1)}} \left(\frac{\dot{\sigma} b}{2v_0 G} \right)^{\frac{(2m+3)}{(m+1)(2m+1)}} \tag{40}$$

Using Equations (7)~(10) and the following approximation for $m = 10$–30 (steel) [16], Equation (42) was obtained.

$$\frac{2m+1}{(m+2)(2m+1)} \approx \frac{1}{m+2} \approx \frac{1}{m+1} \approx \frac{1}{m} \tag{41}$$

$$\sigma_Y = 2\tau_{00} \left\{ \frac{A^*}{1.4} \left(\frac{H_k}{4kT} \right)^{1.45} \frac{1}{\rho_0^* \Lambda b^2} \left(\frac{\tau_{00}}{G} \right)^{-1.9} \right\}^{\frac{2kT}{H_k}} \left(\frac{\dot{\sigma} b}{2A_1 G} \right)^{\frac{4kT}{H_k}} \tag{42}$$

As material constants, the following physically reasonable and almost equal values to those used for the analysis of the delay time of yielding, $\rho_0^* \Lambda = 1.75 \times 10^6 / m^2$ [16], $G = 79.38$ GPa, $\tau_0^* = 176.0$ MPa [18], and $m = 10$ [16] were taken at room temperature, with $b = 3 \times 10^{-10}$ m selected. Before yielding, since $\tau_a = G\gamma_a$ was almost satisfied, A^* was considered to be almost equal to one.

A comparison of Equation (42) with experimental data is shown in Figure 19 [17]. Equation (42) was found to agree well with experimental data [19]. Furthermore, this equation was in good agreement with that obtained based on dislocation dynamics theory that defined yielding as occurring when dislocation density takes some critical value [6] as follows.

$$\sigma_Y = 2\tau_{00} \left\{ \frac{1}{1.4} \left(\frac{H_k}{4kT} \right)^{1.45} \frac{\overline{N_0}}{\rho_0^* \Lambda} \right\}^{\frac{4kT}{H_k}} \left(\frac{G\dot{\sigma} b}{2A_1 \tau_{00}^2} \right)^{\frac{4kT}{H_k}} \tag{43}$$

This means that the γ_P criterion is identical to the $\overline{N_0}$ criterion. Furthermore, Equation (42) was found to be in in good qualitative agreement with theoretical results [8] based on Cottrell-Bilby's dislocation release mechanism [7] for dislocation locking by solute atoms such as carbon or nitrogen, as given by Equation (44). This means that the locking mechanism closely connects with the mechanism of dislocation group dynamics as the appropriate mechanism of locking.

$$\sigma_Y = \sigma_0 \left(\frac{t_0}{nkT} \frac{\dot{\sigma}}{\sigma_0} \right)^{nkT} \tag{44}$$

Furthermore, other results obtained based on a viscoplasticity model also showed the same type of relationship between strain rate and yield stress [5].

From the total results mentioned above, it can be seen the criteria of dislocation release from locking by the solute atom [8], critical strain rate [16,17], critical dislocation density [6], and viscoplasticity [5] are closely connected with the sequential processes of plastic yielding. Furthermore, the theoretical results [8] based on Cottrell-Bilby's dislocation release mechanism [7] from locking by solute atom (Equation (44)) are considered to be a starting process of yielding.

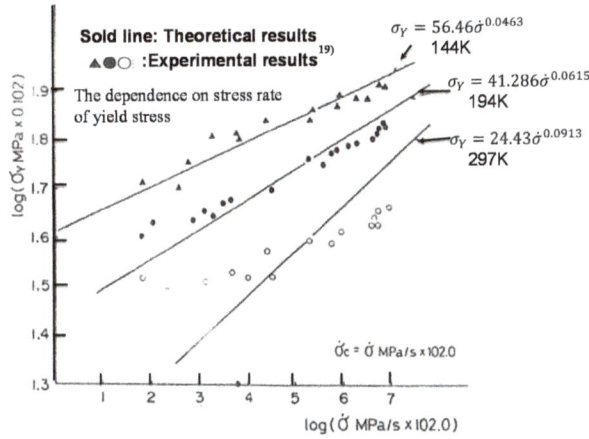

Figure 19. The relationship between upper yield point and constant stress rate based on dislocation dynamics. Solid lines represent the theoretical results [17]. Dotted lines represent experimental results [19].

T. Yokobori also found that the adoption of the friction stress, τ_i, of the motion of dislocation in Equation (40) was very effective in predicting experimental results with the current equation [20]. In this theory for Equation (40), the effect of τ_i on yield stress carries on to lead to Equation (45) [17].

$$\sigma_Y = \sigma_i + 2(\tau_0^* - \tau_i) \left(\frac{m^{1.45}}{1.4} \frac{A^*}{\rho_0^* \Lambda b^2} \right)^{\frac{1}{(2m+1)}} \left(\frac{\tau_0^* - \tau_i}{G} \right)^{-\frac{2(m+1)}{(m+2)(2m+1)}} \left(\frac{\dot{\sigma} b}{2 v_0 G} \right)^{\frac{(2m+3)}{(m+1)(2m+1)}} \tag{45}$$

By adopting $\tau_i = 86.3$ MPa and using Equations (10a)–(10d), Equation (45) was found to well-predict experimental results [19], as shown in Figure 20 [17].

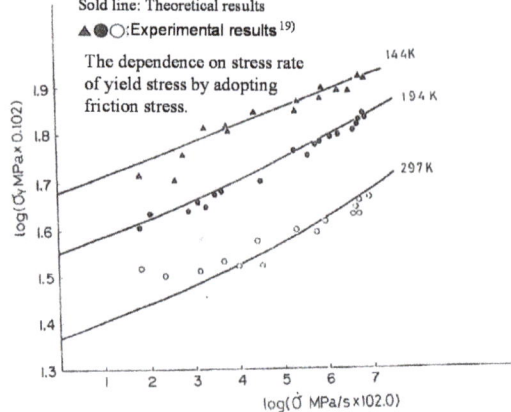

Figure 20. The relationship between the upper yield point and constant stress rate based on dislocation dynamics by accounting for the effect of friction stress of dislocation on upper yield point. Solid lines represent the theoretical results [17]. Dotted lines represent experimental results [19].

The relationship between τ_0^* used for analyses of Figures 17–20 and temperature T is shown in Figure 21. The results were in good agreement with the theoretical relationship given by Equation (10c), thus showing the validity of the method of analysis.

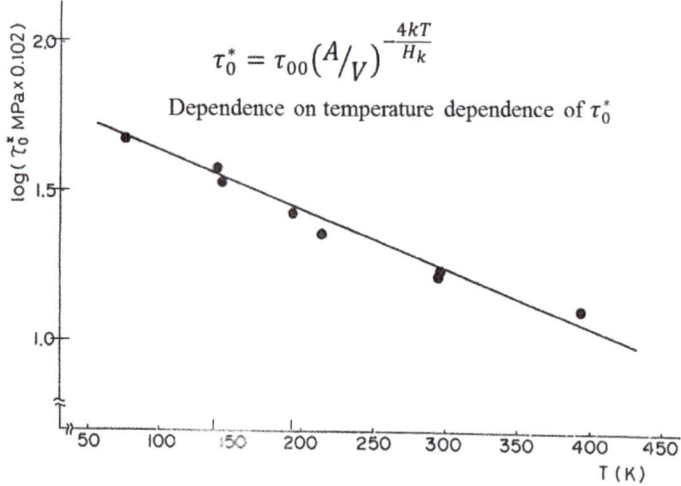

Figure 21. The relationship between τ_0^*, as used for this analysis, and temperature [17].

The Effect of Grain Size and Applied Strain Rate on Yield Stress Based on the Theory of Dislocation Piling Up [2,3,21]

The effect of grain size d on lower yield point was obtained by the following experimental relationship [22,23].

$$\sigma_{l,y} = \sigma_s + \kappa d^{\frac{-1}{2}} \tag{46}$$

where σ_s and κ are material constants that are positive values.

Furthermore, many detailed studies have been conducted on this relationship [24,25].

In this section, using the Equation (25), the calculated relationship between an applied stress, τ, required for a dynamic K_d to take on a critical value over a range in grain size d is shown by the solid line in Figure 22 [3], which is expressed by Equation (47).

$$\tau = \tau_1 + k \frac{1}{\sqrt{d}} \qquad (47)$$

where τ_1 and k are constants. The dashed line is a static solution given by Equation (48).

$$\tau = \frac{K_s}{\sqrt{2\pi d}} \qquad (48)$$

where K_s is static stress intensity factor.

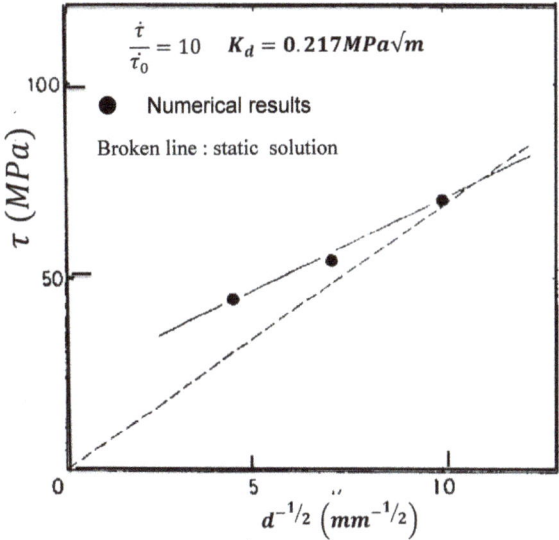

Figure 22. The relationship between yield strength and grain size, d. Solid line represents the results obtained by the analysis of dislocation dynamic pile, as shown by dotted data [3]. Dotted line represents the solution obtained by Equation (48).

It can be seen from Equation (47) that the dynamic yield strength also increases linearly with respect to the inverse square root of grain size. Furthermore, by comparing Equations (47) and (48), the yield stress corresponding to $d^{-\frac{1}{2}} \approx 0$ was found to be higher in dynamic yielding than that in the static case, and this characteristic is in good agreement with the experimental data [26].

Concerning the effect of applied strain rate on yield stress, the rate-determining process of yielding of iron and steel is considered to correspond to the dynamic piling up of emitted dislocations.

The dislocation pile-up at grain boundaries and yielding is considered to occur when K_d, given by Equation (25) to measure the local stress concentration, takes on a critical value. A comparison of results obtained by Equation (25) and experimental data [27] is shown in Figure 23 [3]. In Figure 23, the solid line represents the calculated relation between the applied stress σ required for K_d to take the critical value and applied strain rate that is in good agreement with experimental data [25]. In Figure 23, $\dot{\varepsilon}$ is evaluated by the relationship of $\dot{\sigma} = 2\dot{\tau} = E\dot{\varepsilon}$.

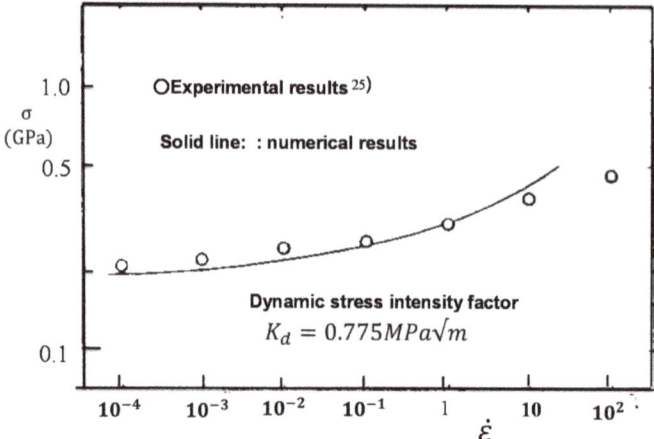

Figure 23. The relationship between strain rate and yield stress of mild steel. Dotted line represents the numerical results [3].

From the total results presented in Section 2.4, it can be seen that the various criteria associated with dislocation release from locking by solute atom [8], critical plastic strain [16,17], critical strain rate or stress rate, critical dislocation density [6], viscoplasticity [5], and dislocation pile-up at a grain boundary [3,4] characterized by local stress intensity factor [3] (K_d) all relate closely to the sequential processes of yielding. Furthermore, the theoretical results [8] based on Cottrell-Bilby's dislocation release mechanism [7] for dislocation locking by solute atoms is considered to be a starting process of yielding.

In addition to the relationship between yield stress and grain size, a theoretical relationship between the yield stress and temperature was derived based on dislocation mechanics [28,29]. The present description closely connects with the results given in Figures 18 and 20 [17].

2.5. Application to Problem of Creep

Previous descriptions of the creep rate have been dominated by the use of equations based on the properties of an isolated dislocation [30].

In this section, instead of using the velocity of an isolated equation, the results of dislocation group dynamics associated with emission from a stressed source under constant stress condition were adopted, and a creep rate dominated by the grouped dislocation mechanism was formulated.

The maximum radius of dislocation loop is given by Equations (49) and (50) [30].

$$L = \sqrt{\frac{2(\tau_a - \tau_i)}{Gb\rho_0^*}} \tag{49}$$

$$\Lambda^* = 3L \tag{50}$$

By using Equations (27)–(29) and Equation (32), the creep rate, $\dot{\gamma}$, can be given by Equation (51).

$$\dot{\gamma} = A_1(\tau_a - \tau_i)^\delta \tag{51}$$

In (51), τ_a in Equation (32) is replaced by $(\tau_a - \tau_i)$.

$$\delta = \frac{2m^2 + \frac{9}{2}m + 2}{m + 2} \tag{52}$$

$$A_1 = 3\sqrt{\frac{2\rho_0^* b}{G}} \times 2.45 m^{-0.86} \left(\frac{Gb}{v_0 t}\right)^{-\frac{m+1}{m+2}} \times (\tau_0^*)^{-\frac{m(2m+3)}{m+2}} v_0 \qquad (53)$$

Furthermore, when using Equations (7)–(10), the creep rate, $\dot{\gamma}$ is given by Equation (54) as an equation of a thermally-activated process.

$$\dot{\gamma} = C_1 (\tau_a - \tau_i)^{\alpha_1} \exp\left(-\frac{Q_1}{kT}\right) \qquad (54)$$

where

$$C_1 = 3\sqrt{\frac{2\rho_0^* b}{G}} \times 2.45 m^{-0.86} A_1 \left(\frac{Gb}{A_1 t}\right)^{-\frac{m+1}{m+2}} \times (G)^{-\frac{m(2m+3)}{m+2}} \qquad (55)$$

$$\alpha_1 = \frac{2m^2 + \frac{9}{2}m + 2}{m+2} \qquad (56)$$

$$Q_1 = \frac{2m+3}{4(m+2)} H_k \ln \frac{\tau_{00}}{G} \qquad (57)$$

Equation (54) is in good agreement with a pioneering experimental equation given for the creep rate [31]. Since $m = 1$ is valid for Zn [13], α_1 is 2.83 in Equation (54), which is in good agreement with experimental data [31]. Thus, Equation (54) is a theoretical equation of a creep rate dominated by a dislocation mechanism that incorporates the effect of dislocation dynamics corresponding to dislocations being emitted from a stressed source.

2.6. Application to the Problem of Fatigue Crack Growth [32,33]

The fatigue crack growth rate da/dN for a crack blunting and re-sharpening model [34] is approximately equal to $\frac{1}{2}U$, as shown in Figure 24 [32,33] and given by Equation (58).

$$\frac{da}{dN} \cong \frac{1}{2}U = nb \qquad (58)$$

where U is the crack opening displacement caused by dislocation emission from a crack tip and is equal to 2nb, b is the Burgers vector, and n is the number of dislocations emitted from a crack tip, as shown in Figure 24.

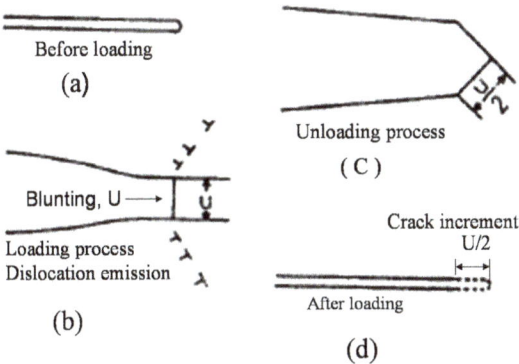

Figure 24. Blunting and re-sharpening model of fatigue crack growth. (**a**) Before fatigue load cycle, (**b**) maximum load of the fatigue cycle, (**c**) unloading process of the fatigue cycle, and (**d**) complete unloading of the fatigue cycle. The crack increment of U/2 (crack opening displacement/2) is caused by dislocation emission.

Substituting Equations (6)–(10) into Equation (58) allows for the fatigue crack growth rate to be expressed in terms of an apparent single thermal activated process that is given by Equations (59a) and (59b) [32,33].

$$\frac{da}{dN} = b\gamma(m)\left(\frac{4fb}{v_0}\right)^{-\frac{m+1}{m+2}} \left(\frac{\tau_0^*}{G}\right)^{-\frac{m(m;1)}{m+2}} \left(\frac{\Delta K_1}{\sqrt{\varepsilon}G}\right)^{\frac{(m+1)^2}{m+2}} \tag{59a}$$

$$= bA^*(4f)^{-\frac{m+1}{m+2}} \left(\frac{\Delta K_1}{\sqrt{\varepsilon}G}\right)^{\frac{m+1}{m+2}} \exp\left\{-\frac{\left(\frac{m+1}{m+2}\right)H_k ln\left(\frac{\tau_{00}\sqrt{\varepsilon}}{\Delta K_1}\right)}{4kT}\right\} \tag{59b}$$

where ΔK_1 is the stress intensity factor, f is load frequency, and ε is the local distance from a crack tip, e.g., 1.5×10^{-4} mm. Local stress around a crack tip is characterized given by $\tau_a = \frac{\Delta K_1}{\sqrt{\varepsilon}}$.

$$A^* = 1.4m^{-1.45}\left(\frac{b}{A_1}\right)^{-\frac{m+1}{m+2}}$$

Equations (59a) and (59b) can be written as Equation (60).

$$ln\frac{da}{dN} = ln\left(\frac{A_2}{f^\lambda}\right) + b_0 ln\Delta K_1 - \frac{U_2 - a_2 ln\Delta K_1}{kT} \tag{60}$$

where A_2, a_2, and b_0 are material constants.

Equations (59a) and (59b) can be expressed as:

$$\frac{da}{dN} = B_1 \Delta K_1^\delta \tag{61}$$

Which is the well-known experimental Equation by Paris [35], where $\delta = \frac{(m+1)^2}{(m+2)}$.

The experimental relationship between $ln\left(\frac{da}{dN}\right)$ and $1/T$ is shown in Figure 25 [36] with the parameter of stress intensity factor amplitude, ΔK_1.

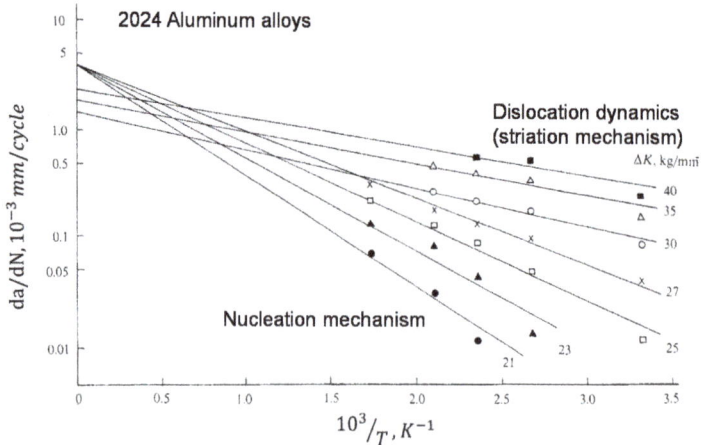

Figure 25. The thermally activated relationship between da/dN (fatigue crack growth rate) and the inverse value of absolute temperature, $\frac{10^3}{T}$, for 2024 aluminum alloys [36].

Figure 25 shows that these relationship were found to hold for the thermally activated process in the range of higher values of ΔK_1, in that the intercept values of the straight line of $ln\left(\frac{da}{dN}\right)$ with the coordinate axis at $1/T = 0$ were an approximately linear function of $\ln(\Delta K_1)$, which is in good agreement with Equation (60), that is, $\ln\left(\frac{A_2}{f^\lambda}\right) + b_0 \ln \Delta K_1$ [32,33].

On the other hand, in the range of lower values of ΔK_1, the intercept value is constant with stress intensity factor amplitude, ΔK_1, which is in good agreement with the model based on micro crack nucleation at the crack tip [37], given by Equation (62).

$$\ln \frac{da}{dN} = \ln\left(\frac{A_1}{f}\right) - \frac{U_1 - a_1 \ln \Delta K_1}{kT} \qquad (62)$$

where A_1 and a_1 are material constants. The same experimental tendencies were also found in the relationship between $\ln\left(\frac{da}{dN}\right)$ and $1/T$ for stainless steel [38]. Furthermore, Equation (60) was found to be in good agreement with the experimental relationship between $\ln\left(\frac{da}{dN}\right)$ and $1/T$ for high strength steel at low temperatures [39].

Experimental data showed that da/dN is proportional to $f^{-\lambda}$, and λ experimentally takes values from 0.1 to 0.2 for steel [40] and 0.1 to 0.5 for aluminum alloys [41].

For the case of an elastic–plastic crack, the local stress around a crack tip is written by Equation (63) [42].

$$\sigma_l = f(\beta)\sigma_{cy}\left(\frac{\Delta K_1}{\sigma_{cy}\sqrt{\varepsilon}}\right)^{\frac{2\beta}{1+\beta}} \qquad (63)$$

where σ_{cy}, β, and $f(\beta)$ are the initial yield stress in cyclic straining, cyclic strain hardening exponent, and some function of β, respectively.

Therefore, by comparing Equations (63), (61), and (59a), da/dN can be given by the following equation [33].

$$\frac{da}{dN} \propto \left(\frac{\Delta K_1}{\sqrt{2\varepsilon\sigma_{cy}}}\right)^{\frac{2\beta}{1+\beta}\frac{(m+1)^2}{m+2}} \qquad (64)$$

Furthermore, for the effect of multiple slip lines and strain hardening under cyclic loading, Equation (64) can be rewritten as Equation (65) [33].

$$\frac{da}{dN} \propto \left(\frac{\Delta K_1}{\sqrt{2\varepsilon\sigma_{cy}}}\right)^{\frac{2\beta}{1+\beta}\frac{(m+1)^2}{m+2}+\frac{1}{1+\beta}} \qquad (65)$$

For $\beta = 0.08$–0.3 and $m = 4$–10, which are reasonable values for steel and aluminum alloys, the power exponent becomes $\delta = 2.0$–5.0, which are also experimentally reasonable values.

In Equation (61), the following equation can be seen from Equation (59a).

$$B_1 = B/\left(\sqrt{\varepsilon G}\right)^\delta \qquad (66)$$

From Equation (66), Equation (67) can be obtained [43].

$$\ln B_1 = \ln B - \delta \cdot \ln\left(\sqrt{\varepsilon G}\right) \qquad (67)$$

Equation (67) was found to be in good agreement with the experimental relationship [43] between B_1 and δ, as shown in Figure 26 [43].

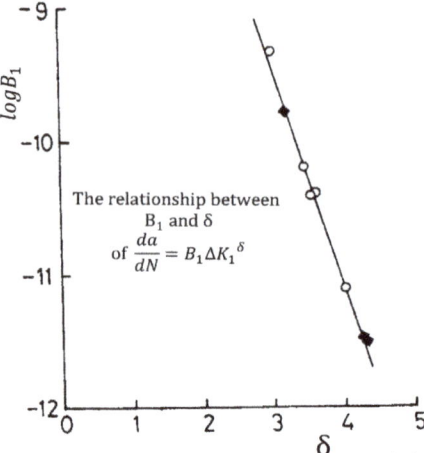

Figure 26. The experimental relationship between B_1 and δ in Equation (67) [43].

3. Concluding Remarks and Future Problem

Analyses of discrete dislocation dynamics and emission from a stressed source were conducted. The results obtained by these analyses enabled us to link various dynamical effects, such as load frequency and temperature, on the yield stress, dislocation creep rate, and fatigue crack growth rate with the experimental results of macroscopic phenomenon and to also link them with theoretical results obtained by the concept of static, continuously distributed infinitesimal dislocations for the equilibrium state under low strain or stress rate conditions.

This will be useful as a holistic research approach relating to the time scale—e.g., ranging from results under high strain rate condition to those under static or low strain rate conditions—and the space scale—e.g., ranging from meso-scale and macro-scale mechanics—that is, from the scale of dislocation groups dynamics to fracture mechanics.

To establish a perfect link of mesoscopic mechanics with macro mechanics and for practical applications to engineering structures, further nonlinear interactive treatments will be necessary, e.g., effects of vacancy diffusion, different multiaxial stress in structures, and different scales of grain boundary influences on dislocation group dynamics. For these study fields, the establishment of inter disciplinary science between material science and structural engineering coupled with computational mechanics is needed as one of future research problems involving the strength of materials.

Detailed research on the effects of grain size and temperature on the yield stress has been systematically conducted, and many innovative results have been obtained [28,29].

The proposed research approach mentioned in this article will enable us to link mesoscopic mechanical factors with macro-scale engineering results [28,29].

Concerning the problems of nano-scale fracturing and plasticity, many studies based on the method of 3D discrete dislocation dynamics have already been successfully conducted [1]. These studies would appear to directly connect with physical properties of dislocations and nano-scale fracturing behaviors. The present results should lead to a wider establishment of fracture prediction in the full range from the nano-scale to macro-scale.

Funding: This research received no external funding.

Acknowledgments: I deeply appreciate R.W. Armstrong and A. Zubelewicz for useful discussions. I also deeply appreciate co-authors of my articles referred in this articles for supports of my research.

Conflicts of Interest: The authors declare no conflict of interest.

References

1. Zhou, C.; Biner, S.B.; LeSar, R. Discrete dislocation dynamics simulation of plasticity at small scales. *Acta Metall.* **2010**, *58*, 1565–1577. [CrossRef]
2. Yokobori, T.; Yokobori, A.T., Jr. Physical and Phenomenological Model with Non-Linearity in Ductile fracture and fatigue crack growth. In *Physical Non-Linearity's in Structural Analysis, Proceedings of the IUTAM Symposium, Senlis, France, 27–30 May 1980*; Hult, J., Lemaitre, J., Eds.; Springer: Berlin/Heidelberg, Germany, 1980; pp. 271–286.
3. Yokobori, A.T., Jr.; Yokobori, T.; Nishi, H. Stress Rate and Grain Size Dependence of Dynamic Stress Intensity Factor by Dynamical Piling-up of Dislocations Emitted. In Proceedings of the IUTAM Symposium on MMMHVDF, Tokyo, Japan, 12–15 August 1985; Kawata, K., Shioiri, J., Eds.; Springer: Berlin/Heidelberg, Germany, 1987; pp. 149–164.
4. Suh, N.P.; Lee, R.S. A dislocation model for the delayed yielding phenomenon. *Mater. Sci. Eng.* **1972**, *10*, 269–278. [CrossRef]
5. Zubelewicz, A. Review Mechanical-based transitional viscoplasticity. *Crystals* **1972**, *10*, 212. [CrossRef]
6. Yokobori, A.T., Jr.; Yokobori, T.; Kawasaki, T. Computer simulation of dislocation groups dynamics under applied constant stress and its application to yield problems of mild steel. In *High Velocity Deformation of Solid*; IUTAM Symposium; Kawata, K., Shioiri, J., Eds.; Springer: Berlin/Heidelberg, Germany, 1979; pp. 132–148.
7. Cottrell, A.H.; Bilby, B.A. Dislocation theory of yielding and strain ageing of iron. *Proc. Phys. Soc. Lond. A* **1949**, *62*, 49–62. [CrossRef]
8. Yokobori, T. Delayed yield and strain rate and temperature dependence of yield point in Iron. *J. Appl. Phys.* **1954**, *25*, 593–594. [CrossRef]
9. Yokobori, T.; Yokobori, A.T., Jr.; Kamei, A. Computer simulation of dislocation emission from a stressed source. *Philos. Mag.* **1974**, *30*, 367–378. [CrossRef]
10. Yokobori, A.T., Jr.; Yokobori, T.; Kamei, A. Generalization of computer simulation of dislocation emission under constant rate of stress application. *J. Appl. Phys.* **1975**, *46*, 3720–3724. [CrossRef]
11. Kanninen, M.F.; Rosenfield, A.R. Dynamics of dislocation pile-up formation. *Philos. Mag.* **1969**, *20*, 569–587. [CrossRef]
12. Rosenfield, A.R.; Hahn, G.T. Linear arrays of moving dislocations emitted by a source. In *Dislocation Dynamics*; Rosenfield, A.R., Hahn, G.T., Eds.; Mcgraw-Hill: New York, NY, USA, 1968; pp. 255–273.
13. Turner, A.P.L.; Vreeland, T. The effect of stress and temperature on the velocity of dislocations in pure iron mono crystals. *Acta Metall.* **1970**, *18*, 1225–1235. [CrossRef]
14. Seeger, A. On the Theory of the Low-temperature Internal friction Peak Observed in Metals. *Philos. Mag.* **1956**, *1*, 651–662. [CrossRef]
15. Johnston, W.G. Yield points and delay times in single crystals. *J. Appl. Phys.* **1962**, *33*, 2716. [CrossRef]
16. Hahn, G.T. A model for yielding with special reference to the yield-point phenomenon of iron and related bcc metals. *Acta Metall.* **1962**, *10*, 727–738. [CrossRef]
17. Yokobori, T., Jr.; Yokobori, T.; Sakata, H. Derivation of Plastic Strain Rate Formula Based on Dislocation Groups Dynamics and Its Application to Yield Problems. *Jpn. Soc. Mech. Eng.* **1984**, *50*, 654–660. (In Japanese) [CrossRef]
18. Stein, D.F.; Low, J.R., Jr. Mobility of edge dislocation in silicon-iron crystals. *J. Appl. Phys.* **1960**, *31*, 362. [CrossRef]
19. Hendrickson, J.A.; Wood, D.S.; Clark, D.S. Prediction of transition in a notched bar impact test. *Trans. Am. Soc. Met.* **1959**, *51*, 629.
20. Yokobori, T. *Zairyo Kyoudogaku*, 2nd ed.; Gihodo Pub.: Tokyo, Japan, 1955; Iwanami Pub.: Tokyo, Japan, 1974.
21. Gerstle, F.P.; Dvorak, G.J. Dynamics formation and release of a dislocation pile-up against a viscous obstacle. *Philos. Mag.* **1974**, *29*, 1337–1346. [CrossRef]
22. Sylwestrowicz, W.; Hall, E.O. The deformation and ageing of mild steel. *Proc. R. Soc. Lond. B* **1951**, *64*, 405–502. [CrossRef]
23. Petch, N.J. The cleavage strength of poly-crystals. *J. Iron Steel Inst.* **1953**, *174*, 25–28.
24. Armstrong, R.W. The (cleavage) strength of pre-cracked polycrystals. *Eng. Fract. Mech.* **1987**, *28*, 529–538. [CrossRef]

25. Armstrong, R.W. Material grain size and crack size influences on cleavage fracturing. *Philos. Trans. R. Soc. A* **2015**, *373*, 20140124. [CrossRef]
26. Cambell, J.D.; Harding, J. *Response of Metals to High Velocity Deformation*; Inderscience Publishers: New York, NY, USA, 1961; p. 51.
27. Manjoine, M.J. Influence of rate of strain and Temperature on yield stress of Mild SteelTrans. *J. Appl. Mech. Trans. ASME* **1944**, *66*, A-211.
28. Zerilli, F.J.; Armstrong, R.W. Dislocation mechanics based constitutive relations for material dynamics calculations. *J. Appl. Phys.* **1987**, *61*, 1816–1825. [CrossRef]
29. Armstrong, R.W. Takeo Yokobori and Micro-to Macro-Fracturing of poly-crystals. *Strengh Fract. Complex. Int. J.* **2020**, *12*, 79–88. [CrossRef]
30. Weertman, J. Steady-state creep through dislocation climb. *J. Appl. Phys.* **1957**, *28*, 362. [CrossRef]
31. Gilman, J.J. Plastic Anisotropy of Zinc Monocrystals. *J. Met.* **1956**, *8*, 1326–1336. [CrossRef]
32. Yokobori, T.; Yokobori, A.T., Jr.; Kamei, A. Dislocation dynamics theory for fatigue crack growth. *Int. J. Fract.* **1975**, *11*, 781–788. [CrossRef]
33. Yokobori, T.; Konosu, S.; Yokobori, A.T., Jr. Micro and Macro Fracture Mechanics Approach to Brittle Fracture and fatigue crack Growth. *Fracture* **1977**, *1*, 665–682.
34. Laird, C.; Smith, G.C. Crack propagation in high stress fatigue. *Philos. Mag.* **1962**, *7*, 847–857. [CrossRef]
35. Paris, P. Fatigue: An interdisciplinary approach. In Proceedings of the 10th Sagamore Army Materials Research Conference, Sagamore Conference Center, Raquette Lake, NY, USA, 13–16 August 1963; Burk, J.J., Reed, N.L., Eds.; Syracuse University Press: Syracuse, NY, USA, 1964; p. 107.
36. Yokobori, T.; Aizawa, T. The Influence of Temperature and fatigue Crack Propagation rate of Aluminum Alloy. *Int. J. Fracture* **1973**, *9*, 489–491.
37. Yokobori, T. A Kinetic Approach to Fatigue Crack Propagation. In *Physics of Strength and Plasticity*; The Orowan Anniversary Volume; Argon, A.S., Ed.; MIT Press: Cambridge, MA, USA, 1969; pp. 327–338.
38. Yokobori, T.; Aizawa, T. The influence of temperature and stress intensity factor upon the fatigue crack propagation rate and striation spacing of 304 stainless steel. *J. Jpn. Inst. Met.* **1975**, *39*, 1003–1010. (In Japanese) [CrossRef]
39. Kawasaki, T.; Nakanishi, S.; Sawaki, Y.; Hatanaka, K.; Yokobori, T. Fracture Toughness and Fatigue Crack Propagation in High Strength Steel at Low Temperature. *Jpn. Soc. Mech. Eng.* **1975**, *41*, 3324–3331. (In Japanese)
40. Yokobori, T.; Sato, K. The Effect of Frequency on fatigue Crack Propagation Rate and Striation Spacing in 2024-T3 Aluminum Alloy and SM-50 Steel. *Eng. Fract. Mech.* **1976**, *8*, 81–88.
41. Hartman, A.; Schijve, J. The effect of environment and load frequency on the crack propagation law for macro fatigue crack growth in Aluminum alloys. *Eng. Fract. Mech.* **1970**, *1*, 615–631. [CrossRef]
42. Rice, J.R. Stress due to a Sharp Notch in a Work-Hardening elastic-Plastic Material Loaded by Longitudinal shear. *J. Appl. Mech.* **1987**, *34*, 287–298. [CrossRef]
43. Yokobori, T. A Critical evaluation of mathematical equations for fatigue crack growth with special reference to ferritic grain size and monotonic yield strength dependence. In *Fatigue Mechanisms*; Fong, J., Ed.; ASTMSTP: West Conshohocken, PA, USA; Volume 675, pp. 683–706.

© 2020 by the author. Licensee MDPI, Basel, Switzerland. This article is an open access article distributed under the terms and conditions of the Creative Commons Attribution (CC BY) license (http://creativecommons.org/licenses/by/4.0/).

MDPI
St. Alban-Anlage 66
4052 Basel
Switzerland
Tel. +41 61 683 77 34
Fax +41 61 302 89 18
www.mdpi.com

Metals Editorial Office
E-mail: metals@mdpi.com
www.mdpi.com/journal/metals

www.ingramcontent.com/pod-product-compliance
Lightning Source LLC
LaVergne TN
LVHW070703100526
838202LV00013B/1021